To Stem the Tide

Effective State Marine Fisheries Management

The Task Force on Effective State Marine Fisheries Management Programs was sponsored by the Council of State Governments and financially supported by the National Marine Fisheries Service which is a branch of the National Oceanic and Atmospheric Administration within the U.S. Department of Commerce. The contents of this publication are the results of the Task Force effort and do not necessarily reflect the views of the U.S. Department of Commerce.

Editors
Ralph J. Marcelli
Robert D. Matthews

THE COUNCIL OF STATE GOVERNMENTS
Lexington, Kentucky

Published September 1975 by
The Council of State Governments
Iron Works Pike, Lexington, Kentucky 40511

Printed in the United States of America

You did not kill the fish only to keep alive and to sell for food, he thought. You killed him for pride and because you are a fisherman. You loved him when he was alive and you loved him after. If you love him, it is not a sin to kill him. Or is it more?

Ernest Hemingway
The Old Man and the Sea

ISBN 0-87292-014-3
Price: $3.00
RM—566

Contents

The National Task Force on Effective State Marine Fisheries Management Programs

Honorable Bernard C. Smith, New York Senate, Chairman
Dr. Thomas E. Kruse, Director, Fish Commission of Oregon, Vice-Chairman
Honorable Thomas J. Anderson, Michigan House of Representatives
Frank Grice, Director of Marine Fisheries, Department of Wildlife, Fisheries and Recreational Vehicles, Commonwealth of Massachusetts
E. J. Huizer, Deputy Commissioner, Department of Fish and Game, State of Alaska
Tom Linton, Director, Office of Marine Affairs, State of North Carolina
Ed McCoy, Commissioner, Division of Marine Fisheries, Department of Natural and Economic Resources, State of North Carolina
Honorable John A. Nejedly, Chairman, Committee on Natural Resources and Wildlife, California Senate
Honorable Curtis Peterson, Florida Senate
Robert Rubelmann, Fisheries Administration, State of Maryland
Honorable Sanford R. Steckler, Mississippi Senate
Robert D. Testa, Consultant to Senate Committee on Natural Resources and Wildlife, State of California
Honorable Tom Uher, Texas House of Representatives
Dr. W. Mason Lawrence, *Project Director*
Robert D. Matthews, *Staff Assistant*

James M. Ridenour, *Project Coordinator*

Acknowledgments

As with any effort of this nature, special thanks must be given to all those individuals who gave so unselfishly of their time in assisting the Task Force to complete its duties. To list all those individuals who were involved would take a great deal of space; however, there are certain individuals who deserve special recognition: Dr. W. Mason Lawrence; James Ridenour, Robert D. Matthews, and Mrs. Brenda Bivins of the Council of State Governments; and Richard H. Schaefer, Robert Hutton, Paul Kiefer, and Denton Moore of the National Marine Fisheries Service.

Honorable Bernard C. Smith
Dr. Thomas E. Kruse

Foreword

In July 1974, the Council of State Governments established a National Task Force on Effective State Marine Fisheries Management Programs, whose primary responsibility was to develop suggested state legislation in this area. The Marine Fisheries Management Act was brought to the attention of the National Conference on Effective Management of Marine Fisheries at Hyannis, Massachusetts, in June 1975. Presentations before that conference have been included in the appendix to this report to serve as an informational and substantive contribution to the entire project.

A detailed analysis of specific sections within the act and a series of recommendations precede the suggested state legislation. Not only do the recommendations pertain to various sections of the suggested legislation, but also to certain areas which the Task Force felt could not be adequately addressed in the legislation. Various alternatives are set forth dealing with the complex issue of fisheries resource management.

Appreciation is expressed to the Task Force members for the time and effort they have devoted to this project, and to the National Marine Fisheries Service for its cooperation and support.

Brevard Crihfield
Executive Director
The Council of State Governments

Lexington, Kentucky
September 1975

1 | Introduction

In June 1974, the National Marine Fisheries Service contracted with the Council of State Governments to produce suggested state legislation for effective management of marine fisheries. The Council set up a National Task Force, composed of elected and appointed state officials with a variety of interests and backgrounds in fisheries issues, to develop this legislation. Eleven coastal States were represented on the Task Force. During an eight-month period the Task Force met four times to discuss the various issues and drafts which eventually led to the suggested state legislation included in this report and presented at the National Conference on Effective Management of Marine Fisheries held at Hyannis, Massachusetts, on June 24-25, 1975.

In the course of developing the suggested legislation, the staff of the Task Force examined the marine fisheries laws of the 22 coastal States, eight Great Lakes States, American Samoa, Guam, Puerto Rico, and the Virgin Islands, and conducted interviews with the heads of the state fisheries agencies and representatives of the recreational and commercial fishing and environmental groups in 21 States. In addition to the suggested "Marine Fisheries Management Act," this report contains an overview of the major provisions of the suggested legislation, more detailed background information on the most important of these provisions, and the recommendations of the National Task Force.

The suggested state legislation which is included in this report should not be considered a uniform law. Two of the primary reasons for this are: (1) due to the uniqueness of each coastal and Great Lakes State, certain portions of the legislation may not be relevant to a particular State's need. Therefore, careful consideration must be given to the applicability of each section; and (2) since the issue is of such a broad nature, it was felt that by addressing only the major components of the issue a foundation could be created for each State to build upon or improve its fisheries resource management program.

2 | Overview of the Legislation

In the process of developing suggested marine fisheries legislation, major provisions of the act presented certain problems: (1) organization and operation of marine fisheries agencies, (2) intergovernmental relations and fisheries management, (3) statistical information, and (4) controlling entry into fisheries.

The Task Force felt that these areas required further amplification so that the reader could better understand why certain points were addressed in specific clauses and gain a better insight into the many issues involved in state marine fisheries management programs.

Organization and Operation of Marine Fisheries Agencies

Organization

State marine fisheries agencies are organized along the lines of, or included within one of, two basic forms of organization: commissions or departments. A commission-type organization usually consists of five to nine members, appointed by the Governor with the consent of the Senate. The terms of the members are staggered so that following initial appointment a majority of the members will not come up for appointment during a single gubernatorial term. There is often a provision requiring some type of bipartisan balance in the membership. Members either receive a daily fee for their services plus expenses or, more commonly, are unsalaried and receive only necessary expenses. The functions normally assigned to a commission are: to employ a director for the agency, establish policy, and approve budgets, programs, and regulations. The duty of the director is to administer the approved policies and programs of the commission and to recommend policies, programs, and other appropriate actions to the commission. The director's responsibilities include the hiring of personnel and the direction of the day-to-day operations of the agency.

The department-type organization is usually headed by a commissioner who is appointed by the Governor with the consent of the Senate and serves at the Governor's pleasure. Usually he is a member of the Governor's cabinet. The commissioner has all the duties which are shared by the commission and director in the commission-type organization.

There are some basic differences between the two types of organizations. The commissions are intended to be nonpartisan or at least bipartisan.

Advocates of the commission-type organization feel that the commission, with its shared decision-making, is more likely to be responsive to the public interest and can withstand the pressure of special interest groups better than a single commissioner. Special arrangements are required, however, for commissions to handle emergencies and other major actions requiring prompt attention.

In the department-type agency, the commissioner is directly responsible to the Governor who appointed him and to carrying out the policies and programs of his administration. Advocates of the department-type agency emphasize that the commissioner, having final authority, can respond immediately to emergencies and to other matters on which quick action is needed.

There is considerable variation among the 30 coastal and Great Lakes States, American Samoa, Guam, Puerto Rico, and the Virgin Islands in the position of the marine fisheries agency in the basic organization. In some States, the marine fisheries agency is an independent organization or a separate department. In other States, the marine fisheries agency may be a major division of a larger natural resources or environmental department. In still other States, it may be a part of a fish and wildlife agency, which in turn is part of a larger agency.

The main effect of the position of the fisheries agency is on its relation to the final decision-making authority. With an independent agency or separate department, responsibility for final decision-making rests with its commission or commissioner. In fisheries agencies which are a division of a larger department or a part of a fish and wildlife agency, there are usually one or more levels of approval between the director and the person or entity making the final decision. In this situation, fisheries matters may have to compete with policies and programs of other resource units in the agency for the time and attention of those persons responsible for approvals and final decisions. The result may be that fisheries matters do not receive the consideration they merit. On the positive side, being a part of a larger resource organization usually affords greater and more direct opportunity for the fisheries agency to participate in the development of policies and programs for related resources, which may affect fisheries, and to receive input into the development of fisheries programs from those responsible for the management of other resources.

It was concluded that both types of organizational arrangements have their advantages and limitations, and either type can operate effectively and efficiently. Hence, there is included in the suggested state legislation two alternatives to Section 4, one establishing a commission-type agency and one establishing a department-type agency.

Operation

With reference to the operation of state fisheries agencies, a major problem faced by many agencies is that their management capabilities are severely constrained by the lack of adequate authority to regulate fisheries within the States' jurisdiction. A few state fisheries agencies have appropriate authority to regulate fisheries, but in most States the fisheries agency's authority to regulate ranges from none to limited authority in a few defined situations. In the latter situation, management regulations are established by legislation. The delay and uncertainty of management by legislation prevents effective action since most State Legislatures meet for only a few months a year and in some States only every other year. The problem is greatly increased when two or more States are involved in the management of a shared resource.

At present there are no overall management programs to protect fish stocks as a whole off the U.S. coasts and assure the continued well-being of these national assets. The anticipated extension of U.S. fisheries jurisdiction to 200 miles from our coasts will give the U.S. its first real opportunity to deal comprehensively with the management problems of its coastal fisheries. For States to participate effectively in the management of these fisheries, they must have the capability to cooperate with adjoining States and the federal government in the development and implementation of unified management plans. Extended jurisdiction will thus make it even more essential that state fisheries agencies have adequate authority to regulate fisheries.

The consensus was that state marine fisheries agencies should have the authority to adopt, modify, and repeal regulations pertaining to the management of marine fisheries resources. It is also important that the state fisheries agencies be given the authority to adopt emergency regulations to be effective on promulgation and to be subject to comment, objection, and public hearing within a reasonable time following promulgation. Such emergency regulations may be needed to protect fish stocks for which a harvest quota has been established, to prevent serious depletion of certain stocks, and to protect public health.

It should be noted that many of the coastal States have outmoded or inappropriate fisheries laws and regulations on their books. Each of these States should take appropriate action to repeal or modify any existing laws or regulations that do not fulfill a valid management purpose.

Sections 6, 7, and 8 of the suggested legislation are designed to provide the state fisheries agency with adequate authority to manage the fisheries and to take emergency action when needed to protect the public interest. These provisions will give a State the necessary flexibility to cooperate with adjoining States and the federal government in the management of shared resources and to respond promptly and effectively to changing needs of management within the State. For those States whose fisheries agencies do not have adequate regulatory authority, these three sections are considered to be the most important in the suggested legislation.

Intergovernmental Relations and Fisheries Management

Marine fisheries resources are no great respecters of artificially set political boundaries. What may be a great natural resource off the coast of Mississippi today may be off the coast of Louisiana tomorrow. Resources are often highly mobile as are the fishermen, both recreational and commercial, who pursue them.

The enactment by different governmental entities of widely divergent laws, rules, and regulations over fishermen seeking the same resource causes confusion and chaos among the users and often has serious implications for the survival of the fisheries stocks. Therefore, effective management of the resources should be based on sound biological, social, and economic objectives irrespective of governmental boundaries. The management emphasis must be resource-wide, no matter the political jurisdiction in which the fisheries stocks currently reside.

The additional complications of organizing to administer a fisheries management program under some form of extended jurisdiction heightens the need to study the States' role in effective fisheries management. The range of management schemes that are suggested in discussions of proper fisheries management varies all the way from limited local control, to complete state control, to complete federal control, or combinations thereof. Among this wide range of options fall a number of management alternatives.

Local Government

The mobility of fisheries stocks generally is such that local governments have not controlled an area of sufficient size to play a critical role in the effective management of the resource. That is not to say that local governments should not have a voice in deciding the kinds of fisheries activities that are taking place off their shores.

Too often local pressure groups with little regard for sound fisheries management principles have been successful in lobbying general laws of local application through the State Legislatures. For example, a district that is largely tourist-oriented might get a bill passed to limit commercial fishing in its area. The opposite might be true in an area largely supportive of commercial fishing interests. These types of laws are often in direct conflict with scientifically established fisheries management data, are difficult to enforce, and may be in conflict with the continuing well-being of the fisheries resource.

Interstate Agreements and Compacts

Although the traditional "states' rights" position runs high among many States, administrators and legislators are becoming more cognizant of the fact that there must be some form of interstate regulation of migratory species in order to achieve sound management objectives. Thus the need for interstate cooperative management mechanisms is receiving considerable attention.

Left: A catch of shrimp.
Bottom: Closed raceway system for shrimp culture utilizing airlifts for circulation and aeration.

Article I, Section 10, of the U.S. Constitution clearly states that "No State shall, without the Consent of Congress. . .enter into any Agreement or Compact with another State or with a foreign power."

Although the Article seems quite clear and specific in its language, the courts have not been clear in their interpretation. There are instances where States have entered into cooperative arrangements among themselves without the approval of the Congress. The answer to the question of whether an agreement must go to the Congress seems to hinge on the elusive term "national interest." If the agreement is construed to impinge upon the federal right to regulate interstate commerce, it most certainly must be approved by the Congress.

Further, not all States have enabling legislation allowing them to enter into cooperative or interstate agreements for fisheries management decisions. In the absence of such statutory language, States have little authority to promulgate sound fisheries management programs for the migratory fisheries stocks.

The formal compact arrangement approved by Congress and signed by the President is not a new way for States to approach the problems of interstate management of a resource. As of 1970 there were 160 compacts in effect and more than 30 additional compacts under consideration. In the fisheries area, the Atlantic States Marine Fisheries Compact received congressional consent in 1942. Two other compacts, representing the Gulf States and the Pacific States, have been formed since that time.

Basically these compacts were formed to promote better utilization of the fisheries resources in their areas. They have been used extensively as a forum for gathering state elected and appointed officials for discussion and debate over the establishment of sound fisheries management practices.

The compacts themselves were not designed to be managers of resources although Amendment I of the Atlantic States Marine Fisheries Compact and Article X of the Gulf States Compact allow for two or more member States, by action of their Legislatures, to designate the commission as a joint regulatory authority for a specific fishery. Maine, New Hampshire, and Massachusetts have taken advantage of Amendment I by entering into an agreement to regulate the northern shrimp fishery.

Based on economic reasons, some state officials indicate a reluctance to enter into such state compact arrangements. Secondly, under the compact arrangement States not only must give up some of their authority to regulate but they must also provide funds to accomplish the objectives.

Uniform State Laws

Another management mechanism available to States in managing their fisheries resources is the adoption of uniform or similar state laws. Even though the suggested state legislation contained in this report is not to be interpreted as a model uniform law, it should be considered as a beginning for developing analogous state laws.

Federal Management

Federal preemption is a matter of increasing concern among the States on many issues. Federal preemption of the States' authority within the three-mile limit of the territorial waters has received little serious consideration from either the States or the federal government. Most constitutional scholars would hold that the federal government could preempt the States in the management of coastal fisheries by using the right to regulate interstate commerce. However, given the present political climate, it seems unlikely that the federal government would decide to preempt the States in this area. Much more likely is some form of expanded federal/state cooperation to manage interstate fisheries stocks.

State/Federal Cooperative Fisheries Management

The state/federal partnership concept for managing fisheries evolved from a recommendation made in 1969 by the President's Commission on Marine Science, Engineering and Research (the Stratton Commission) which suggested that the rehabilitation of domestic fisheries depended upon the elimination of overlapping, and sometimes conflicting, laws and regulations which have tended to hamper even those fisheries which are economically viable. The report called for the development of a new framework based on national objectives and sound scientific data for the management of fisheries resources.

The state/federal mechanism for fostering cooperation and shared decision-making began in mid-1972 with the initiation of cooperative management planning for the American lobster fishery. Since that time, development of management plans has been initiated for five additional fisheries. These are the Dungeness crab fishery on the West Coast, the surf clam in the Middle Atlantic Bight, South Atlantic shrimp, Gulf of Maine shrimp, and Gulf of Mexico menhaden. Two additional fisheries have been nominated by the States for initiating management plans; these are the Alaska king and snow crab, and certain Southern California coastal species.

The possibilities of the United States entering into some arrangement for extended jurisdiction, either through negotiated agreement or unilateral action, seem quite high. Logically the federal government would not assume increasing responsibilities without strong cooperation and assistance from state and local governments.

Under any form of extended jurisdiction, it would be safe to assume that the federal government will be looking to the design of fisheries management policies and programs that will strongly involve state and local governments. If recent patterns of federal/state program design are used, States will most likely be offered financial incentives to help plan and implement a national fisheries management system. The issue of federal preemption of a State in the area of fisheries management will more than likely play a role in the incentives program as it has in many of the environmental protection, transportation, and other programs that have been characterized as national in interest.

These arrangements will most likely be based on a regional approach to fisheries management with the different levels of government playing a role in both policy analysis and program implementation.

Summary

It is evident that if the people of the U.S. are to receive the maximum benefits from their valuable marine fisheries resources, these resources must be properly managed. To achieve such management of fish resources shared by several States or shared among several States and the federal government, each affected State must have the ability to participate effectively in the development and carrying out of the management plans. Section 9 of the suggested legislation makes it the responsibility of the state fisheries agency to cooperate with other States and the federal government to develop integrated management plans for shared fisheries resources. This section also provides for coordination between the fisheries agencies and other state agencies whose activities affect fish resources.

Section 10 empowers a state fisheries agency to enter into reciprocal agreements with an adjoining State for joint management of fisheries in boundary waters. Specifically, it provides for adoption of unified regulations, reciprocity in licensing, and "hot pursuit" of violators of fisheries regulations.

Statistical Information

For a number of years, fisheries management agencies have relied heavily on measurements of the amount of fish caught and the effort required to catch them to estimate the status of fisheries stocks and the impact of fishing on them. There are three major limitations to the data currently being collected. One, the data for a number of fisheries is inadequate or unreliable. Two, for fisheries stocks which are harvested by fishermen from two or more States, the data collected separately by individual States is frequently not compatible and cannot be brought together to give valid measurements of catch and effort. Three, there is very little information being obtained on the recreational catch. This consideration is growing in importance as the recreational fishing pressure increases and the recreational catch makes up a substantial part of the total catch or exceeds the commercial harvest in some fisheries.

The problems of inadequate and unreliable data are largely the result of the States' lack of authority to require reporting of needed information and the lack of funds to carry out an effective data collection and analysis program. The desire of fishermen to maintain secrecy as to gear used, fishing locations, income, expenditures, and other practical matters of actual operation is an associated problem. Confidentiality of data is an extremely sensitive issue to fishermen whose personal statistics are involved in a data system.

Some Effective Statistical Programs

California, Oregon, and Washington have fisheries statistics collection systems that are relatively advanced and efficient for each State's internal need. However, with the albacore fishery which is pursued by fishermen from California, Oregon, and Washington, the separate systems were not sufficiently compatible to provide reliable coastwide information on catch, effort, and value. These States have developed and are implementing a coastwide data system for the albacore fishery. A problem that had to be overcome in developing the coastwide system was the limitation in state statutes on the interchange and publication of fishery statistics. The system developed by these States and their experience in implementing it will be helpful in developing data collection and analysis systems for other stocks shared by two or more jurisdictions.

Catch and Effort Data

The lack of catch and effort data on the recreational fisheries is due to the large costs involved in obtaining reliable estimates on the catch of more than 10 million marine recreational fishermen. The National Marine Fisheries Service has instituted a project to obtain this data on a regional basis. Two major steps are involved. First, there is a sampling of the population in a region to determine which areas are engaged in marine recreational fishing (this is necessary since marine recreational fishermen are identified in only the few States that require marine recreational fishing licenses) and to determine the percentage of the population that are marine recreational fishermen. Second, there must be a sampling of the identified marine recreational fishermen to obtain catch and effort data. Present plans are to cover the coasts of the continental United States in a three-year cycle, which would result in revising and updating of the information for each region every three years. Intensification and refinement of this system, with participation by state fisheries agencies as appropriate and possible, could provide information on the recreational fisheries needed for effective management of coastal fisheries. The cost of collecting statistics on the recreational fisheries could be substantially decreased and the efficiency of the collection system improved if each coastal State would institute a licensing system for marine recreational fishermen.

Information on the abundance, distribution, and condition of fish stocks and the effects of various fishing levels and environmental changes on stock abundance is essential for the effective management of fisheries resources. Much of the needed information can be supplied by the fishermen through properly designed reporting systems. Statistics that are as complete and accurate as practical are needed for management of domestic fisheries resources and as a basis for international negotiations on fish stocks shared with other nations.

Section 11 provides the legal basis for obtaining catch statistics on commercial and recreational fisheries. It provides flexibility that will permit a State to cooperate with other States in obtaining integrated catch statistics on a regional or coastal basis. The provisions relating to confidentiality are designed to protect the personal statistics of an individual fisherman or vessel without imposing unworkable limitations on the use of the data for management purposes.

Aquaculture

Aquaculture, the culture and husbandry of aquatic organisms, has the potential of increasing the supply of certain aquatic species as the maximum sustainable yields of wild stocks are approached or exceeded. Public aquaculture, rearing species for stocking public waters, has been a historic function of the States and in some cases the federal government. Private fish farming in the United States has included few species. However, it already provides over 50 percent of our catfish, 40 percent of our oysters, 20 percent of our salmon, and nearly all of our trout. Additional species produced in smaller quantities include clams, marine shrimp, freshwater prawns, crawfish, and marine plants.

Expansion of aquaculture will be needed to provide adequate quantities of various fisheries products to meet projected U.S. needs. This will require several actions at the state level, including the following:

1. Establishing a state policy to encourage commercial aquaculture as a method of increasing supplies of certain fisheries products;
2. Recognizing, in state shorelines management or coastal zoning plans, the necessity to provide space for commercial aquaculture in bays, estuaries, and near-shore coastal areas;
3. Providing, in river basin plans and in state water quality standards, for adequate supplies of high-quality water needed for aquaculture of freshwater species and clean sea water for culture of marine species;
4. Establishing a simplified system for obtaining the various permits and licenses needed for commercial aquaculture;
5. Conducting research and development programs to provide biological and technological information needed for development of public and private aquaculture;
6. Assisting the private aquaculture industry in solving long-range problems, such as genetic improvement of stocks and disease control and by taking actions in resource emergencies which are beyond the capabilities of the industry; and
7. Providing information and advisory services to fish and shellfish farmers to the extent that those services are provided to fishermen.

Floating salmon farm in Puget Sound, Washington. Annual production capacity is approximately 750,000 pounds of pan-sized salmon.

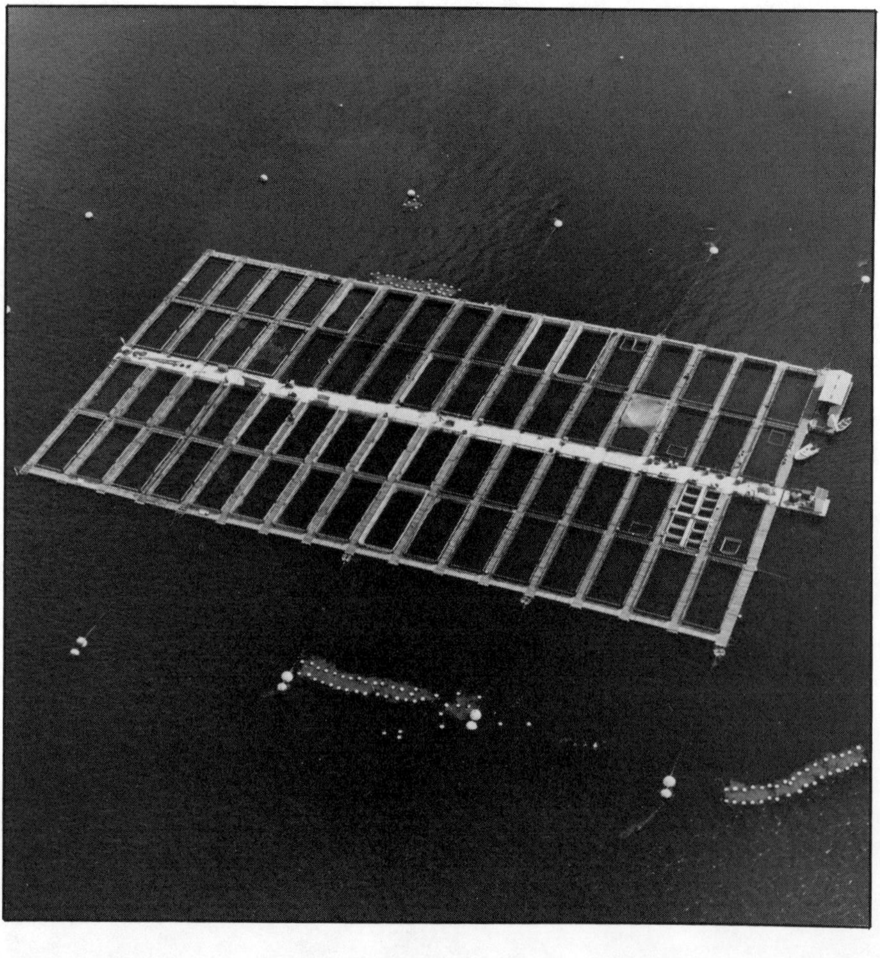

Ocean Ranching

Ocean ranching of homing fishes such as salmon is a special type of aquaculture, which includes the rearing of anadromous species to smolt size in private facilities, releasing them so they can forage in the ocean pastures, and the subsequent harvest of returning adults. This concept, which requires a low input of energy, is appealing as a method of increasing the commercial supply of salmonids. In addition, ocean ranching could increase public recreational and commercial fishing opportunities, since the privately reared fish would mingle at sea with those which result from reproduction of wild stocks. However, resource management would be complicated by the need to insure protection of wild stocks and the public interest, while maintaining the economic incentive for private salmon farming.

Sections 12 and 13 of the suggested state legislation provide statutory bases for a State to permit, encourage, and regulate appropriate aquacultural operations.

Controlling Entry into Fisheries

Since fish are available as a common property for all to harvest, a successful and profitable fishery attracts additional fishermen as long as there is any profit, however small, to be made. Generally the process of free entry into fisheries not only reduces profits to a marginal level but also leads to depletion of the resource. In other renewable resource industries where there is ownership of the resource, the free market tends to be self-regulating. Depletion does not occur because the owners of the resources can respond to the high single scarcity by investing in new productive capacity.

In fisheries, however, the lack of ownership of the resource means that the only way fishermen can respond to scarcity is by attempting to harvest still more of the fish. Fishermen generally cannot produce more fish from a limited resource and as a result the free market in fisheries is not self-regulating but has a tendency to be self-destroying.

If the fishery is managed to conserve the resource, the amount of fish which may be caught in one season will be relatively fixed by regulations. Fishermen respond to such regulations with fishing strategies designed to increase their personal share of the fish available by means of bigger and faster boats as in the tuna fishery, or the use of more gear as in the lobster and crab fisheries. This results in overcapitalization, a situation in which the sum of individual efforts to achieve efficiency leads to inefficiency of the total fleet. In overcapitalized fisheries, profits are marginal, the probability of business failure is high, and the long-run prospects for fishermen to make a decent living are not good. Fishing strategies which lead to overcapitalization also tend to defeat the intended conservation effect of gear restrictions, quotas, minimum size, and other regulations.

An Alternative

A solution to this problem is a limited entry program which limits the number of fishing units participating in the fishery in order to (1) create an environment in which technological improvement results in overall gains in fleet efficiency and to society, (2) insure a healthy economic climate in the fishery, and (3) terminate fishing strategies that are at odds with the goal of conserving the resource.

The limited entry concept already has been applied in certain fisheries. Canada has controlled access into the British Columbia salmon fishery for about 6 years. Alaska, Ohio and Michigan are embarked on similar programs, and the State of Washington has passed legislation authorizing limited entry in the salmon fishery.

In a number of fisheries, limited entry is not an appropriate management measure at this time. The costs of applying entry control to these fisheries would not produce equivalent benefits. There are two types of fisheries that will benefit from controlled entry. One is the traditional, well-developed, higher-value fishery. The second consists of those fisheries in which a considerable growth of fishing effort may be expected in the near future.

Several techniques can be employed to limit entry to fisheries, including license limitations, taxes and fees, and fishermen quotas. None of these options is universally applicable. Careful consideration should be given to which one or which combination would be best suited to a particular fishery.

In implementing limited entry, a means should be provided, usually referred to as grandfathering, of assuring special consideration of fishermen currently participating in the fishery. The transition from unrestricted access to limited entry should be accomplished as fairly as possible and the burden of conserving the resource and the vitality of the industry should be borne in a way that minimizes social and economic dislocation.

A suggested controlled entry provision is included as Appendix 1 to the suggested state legislation for the information and possible use of interested States.

3 | Task Force Recommendations

This chapter lists, in brief fashion, the recommendations of the Task Force. The first section relates to legislative recommendations pertaining to the scope and possible uses of the Marine Fisheries Management Act.

During the course of Task Force deliberations, it became evident that there were certain marine fisheries management issues which could not be adequately addressed in the suggested state legislation. However, the Task Force members felt that even though they could not take specific action in certain areas, it would still be useful to state officials if they made basic policy recommendations in those areas not addressed by the suggested legislation.

Legislative Recommendations

1. Each coastal state fisheries agency should compare the statutory basis for its operation with the suggested "Marine Fisheries Management Act" and seek adoption of those sections or parts, modified to conform to the general construction of its statutes, that would improve its capabilities for marine fisheries management.

2. Each coastal state fisheries agency's attention is directed to the most important needs for modification of present statutes, as determined from an examination of existing state statutes and discussions of problem areas in fisheries management with state fisheries agency heads, and to the sections of the suggested "Marine Fisheries Management Act" that address those needs. Among those needs are adequate regulatory authority within the agency—Sections 6 and 8; adequate catch statistics—Section 11; appropriate licensing of commercial and recreational fishing—Section 6; intergovernmental cooperation—Section 9; advisory input from resource users and interested citizens—Section 5; and effective penalty and enforcement deterrents—Section 14.

3. Since the basic provisions of the suggested "Marine Fisheries Management Act," although designed primarily for marine fisheries management, are equally applicable to the management of freshwater fisheries, the directors of inland fisheries agencies should seek the adoption of those parts of the act that would improve their agencies' capabilities for fisheries management.

Seining for shad, an anadromous fish, in Florida.

Policy Recommendations

4. Marine recreational fishermen should be licensed by the States, with funds being used for management of marine recreational fisheries.

Licensing of marine recreational fishermen is needed to provide a means for collection of basic management information on species of recreational interest such as catch and effort statistics. Moreover, state and federal funds available for management of marine recreational fisheries resources are quite limited and are not likely to be increased appreciably from general fund appropriations. The additional funds that would be generated by recreational fishing licenses should help to insure the conservation and enhancement of recreational fishing resources.

5. Additional funds must be made available to both state and federal fisheries agencies to insure sound management of marine fisheries.

Management costs include those for operations, research, regulation, and enforcement. A management system which will restore depleted fish stocks, develop underutilized fisheries, and maintain fisheries at high-yield levels will cost substantially more than present management efforts, even with improved efficiency of operations. In most States present funding of marine fisheries management is inadequate to support the research and to permit the timely collection and analyses of catch and effort statistics which together constitute basic ingredients of sound management programs.

6. The present level of research should be expanded to provide, on a continuing basis, the needed information on fish stocks to be managed.

Improvement in management and the anticipated extension of United States responsibility for management of fisheries resources in the 200-mile zone will require a corresponding expansion of research to provide an adequate body of information for management. Until this is obtained, decisions will have to be made on the basis of available data, with provisions for revision as the data base is extended and improved.

7. Fish stocks should be managed as an entity rather than differently in each jurisdiction in which they occur or through which they may pass.

Management measures may differ in various parts of a stock's range, based on established need for such variations, but they should be related to an overall management plan for the stock.

8. Maximum opportunity should be afforded the States to participate with the federal government in the management of fisheries resources in the 3-mile zone (or 200-mile zone if the fisheries jurisdiction of the United States is extended).

Many of the fish stocks in the extended zone spend a part of the year in the 3-mile territorial sea within state jurisdiction, and in the case of anadromous fish the spawning areas are sometimes hundreds of miles from the ocean. The United States fishermen who fish for these species are subject to jurisdiction of the States. Therefore, it is essential for effective management that there be maximum involvement of the States in the decision-making process.

9. The U.S. Congress should consider the various Indian treaties with the intent of creating, where needed, a single management authority to conserve our Nation's anadromous fishery resources while protecting those rights reserved to Indians by treaties.

In various parts of the United States, serious problems have developed as a result of Indian tribes exercising treaty fishing rights. Court decisions have gradually increased the legal basis for tribal fishing until, at present, Indian rights in many areas supersede the authority of the State to regulate its fishing resources. The result is a management system where there is divided authority on the regulation of a particular fishery. This divided authority is particularly critical in the case of anadromous fish, which are concentrated in inland streams and can be easily overharvested.

10. State fisheries agencies should improve and expand their programs for public information and education to insure successful implementation and operation of fisheries management programs.

The majority of the state fisheries agencies interviewed recognized the vital role of public information and education programs in carrying out effective fisheries management programs and considered the inadequacy of their present education and information efforts as one of the major weaknesses of their agencies' total program. Historically, agencies have experienced great difficulty in securing adequate funding on a continuing basis for information and education programs because of the belief often held by those controlling appropriations that such programs are actually public relations operations designed to enhance the prestige of the agency and its personnel. In fact, however, public understanding of and participation in fisheries management are essential if the fisheries resources, which are common property of the people of the State, are to be managed in the public interest.

4. Marine Fisheries Management Act

Marine Fisheries Management Act

The Problem. Improving and developing effective fisheries management programs is an issue which is or will be of concern to all coastal and Great Lakes States. However, unless actively involved in the fisheries management process, individuals tend not to perceive this to be a problem of major significance. Most citizens are aware of the fishing efforts by Russian or Japanese ships off our coasts. However, the scope of fisheries management problems is much broader than the above-mentioned incident.

The affected States realize that the sea does not offer a limitless supply of fish to meet the food and recreational needs of our country. As with any renewable resource, there exists a potential of it being overexploited. In various States at this time, certain fisheries resources have reached this point. A consequence of this scarcity in the resource has meant that several commercial operations are in danger of economic collapse.

In addition to the problems of commercial fisheries management is the question of recreational demands upon the resource. In various sections of the country, recreational fishermen look upon commercial fishermen as "plunderers" of fisheries resources. They see commercial fishermen as a threat to the continued enjoyment of recreational fishing. Despite these problems and differences of opinion, both groups have come to realize that it is in their best interest to demand better management to enhance the conservation and wise use of the fisheries resources.

Since state governments have been delegated the primary responsibility for fisheries management within their territorial waters, it is necessary that they attempt to manage the fisheries resources in such a manner that an equitable balance is reached between recreational, economic, and social interests, while at the same time conserving, protecting, and developing fisheries resources. Despite this responsibility, the legislative framework under which state agencies operate is at various levels of sophistication from State to State; therefore, it has become evident that a need exists for a legislative and organizational framework that will allow the most efficient realization of fisheries management objectives on both an intrastate and interstate basis.

By suggesting various organizational mechanisms and management techniques, it is believed that a State can choose those means which will enhance the overall objectives of a fisheries program, i.e., improved management and conservation of the fisheries resources.

This draft legislation was developed by the Council of State Governments' Task Force on Effective Fisheries Management Programs.

Suggested Legislation

(Title, enacting clause, etc.)

Section 1. [*Short Title.*] This act may be cited as the Marine Fisheries Management Act.

Section 2. [*Findings and Declaration of Purpose.*]

(a) The Legislature of this State finds:

(1) That fisheries resources of the State make a material contribution to our economy and food supply, as well as a material contribution to the health, recreation, and well-being of our citizens.

(2) That fisheries resources of the State are a living, renewable form of wealth capable of being maintained and greatly increased with proper management, but equally capable of destruction if neglected or unwisely exploited.

(3) That no person has an inherent right to take or possess fisheries resources without the express consent of the State.

(b) It is hereby declared to be the policy of this State to:

(1) Provide an organizational framework that will permit the State to manage more effectively its fisheries resources with the following objectives:

(i) The conservation of the fisheries resources and their habitat to ensure their continued existence.

(ii) The maintenance and enhancement of fisheries resources to support a recreational use where a species is the object of recreational fishing.

(iii) The maintenance and enhancement of fisheries resources to support commercial use consistent with aesthetic, educational, scientific, and recreational uses of such fisheries resources and the utilization of unused resources.

(iv) The management, on a basis of scientific information, of the fisheries resources under the State's jurisdiction, and the participation in the management of other fisheries in which [State] fishermen are engaged, with the objective of optimum utilization.

(2) Encourage citizen participation through advisory councils and otherwise, since decisions concerning the distribution and allocation of fisheries resources have important consequences for all citizens of this State.

(3) Provide for conservation and management measures involving a reciprocal and cooperative nature among States and between States and the federal government, since many species of fisheries resources travel across state and state-federal boundaries.

Section 3. [*Definitions.*] As used in this act:

(1) "Agency" means the [Department of Marine Fisheries] [Marine Fisheries Commission].

(2) "Director" means the Director of the [Department of Marine Fisheries] [Marine Fisheries Commission].

(3) "Fish" means any marine or anadromous animal, or part thereof, excepting mammals and birds.

(4) "Plant" means any plant, and any parts or seeds thereof, existing in the marine or estuarine waters of the State.

(5) "Take" or "Taking" means to harass, harm, pursue, shoot, wound, kill, net, capture, or collect, or to attempt to engage in any such conduct.

(6) "Fishing gear" means any appliance or device, including vessels or equipment for such vessels, intended for or capable of being used to take fish or plants.

(7) "Fisheries resources" means all fish and plants, and the habitat associated therewith.

(8) "Fishery product" means any product which is made wholly or in part from any fish, or plant, or parts thereof.

(9) "Establishment" means the premises, buildings, structures, facilities, and equipment, including vehicles, used in the processing of fish and fishery products.

(10) "Process," "processed," and "processing," means handling, storing, preparing, reducing, manufacturing, preserving, packing, transporting, holding, or selling thereof.

(11) "Optimum utilization" means a use which provides the greatest benefit to the public as determined on the basis of all relevant economic, social, biological, ecological, and environmental factors.

Alternative A

Section 4. [*Marine Fisheries Commission.*]

(a) There is hereby created a Marine Fisheries Commission.

(b) The Marine Fisheries Commission shall:

(1) Assume the powers, duties, and responsibilities of all other departments, divisions, agencies, or commissions which may be involved in the conservation, enforcement procedures, management, or use of fisheries resources.

(2) Appoint a Director of Marine Fisheries who shall be a person with knowledge of, and experience in the requirements for the protection, conservation, and restoration of the marine fisheries resources of this State. He shall serve for an indefinite term, at the pleasure of the commission. He shall not hold any other public office, and shall devote his entire time to the service of the State in the discharge of his official duties. He shall receive the compensation prescribed by state law, and shall be reimbursed for all actual and necessary traveling and other expenses incurred by him in the discharge of his official duties as delegated to him

by the Commission.

(3) Delegate to the director the authority to adopt emergency regulations pursuant to Section 8.

(c) The commission may delegate to the director any other administrative authority and powers and duties granted to, or imposed upon it, by this act.

(d) The Marine Fisheries Commission shall be composed of [ ] members each of whom is a citizen of this State, who shall have the following qualifications:

(1) Demonstrated interest in public affairs.

(2) General familiarity with fisheries resources problems.

Provided, however, that not more than [ ] such members may be registered voters affiliated with the same political party as the Governor making such appointment.

(e) The members of the Marine Fisheries Commission shall be appointed, serve, and be compensated as follows:

(1) Appointed by the Governor, with the advice and consent of the Senate. If a vacancy shall occur in said commission, it shall be filled in the same manner for the unexpired term.

(2) The terms of office shall begin immediately upon appointment and shall be for [ ] years. However, of the members first appointed [ ] shall be appointed for terms which will expire on [ ]; [ ] for terms which will expire on [ ]; and [ ] for terms which will expire on [ ]. Members shall continue in office until their successors are appointed. An appointee shall be eligible for reappointment.

(3) Be removed by the Governor for cause only.

(4) Be entitled to compensation and expenses as provided.

Alternative B

Section 4. [*Department of Marine Fisheries.*]

(a) There is created and established a Department of Marine Fisheries under the direction of a Director of Marine Fisheries appointed by the Governor, with the advice and consent of the Senate.

(b) The director shall assume the powers, duties, and responsibilities of all other departments, divisions, agencies, or commissions which may be involved in the conservation, enforcement procedures, management, or use of fisheries resources.

Section 5. [*Advisory Council.*]

(a) There shall be a Marine Fisheries Advisory Council to the agency. The council shall be composed of a representative number of private citizens who shall be well versed in the conservation and management of fisheries resources. The [director] [commission] shall appoint the members with the advice and consent of the Senate. One member, by a majority vote of the members, shall be chosen chairperson.

(b) The council shall:

(1) Advise the [director] [commission] on policies of the agency and in the planning, development, and institution of agency programs.

(2) Examine, consider, and make recommendations in any matters pertaining to the conservation and management of fisheries resources throughout the State.

Section 6. [*Powers and Duties of* [*Director*] [*Commission*].]

(a) The [director] [commission] has the following powers and duties:

(1) Promulgate rules and regulations relating to the conservation, management, or use of and enforcement procedures for fisheries resources.

(2) Administer and enforce the laws, rules, and regulations relating to fisheries resources.

(3) Hire all necessary employees of the agency, including enforcement personnel, subject to the Personnel Law of the State.

(4) Organize the agency into such bureaus, divisions, or other administrative units as is necessary to carry out the duties of the agency.

(5) Develop and implement a program for the management of fisheries resources in conformance with this act.

(6) Initiate civil or criminal proceedings when it is necessary and proper.

(7) Acquire and hold real property in the name of the State, or any right or interest therein, including, but not limited to, easements or rights of access.

(8) Subject to the approval of the Governor, accept for the State any federal funds apportioned under federal law related to authorized programs of the agency and to do such acts as are necessary for the purpose of carrying out such federal laws; and to accept from any other agency of government, individual, group, or corporation, such funds, gifts, or devises as may be available to carry out the purposes of the agency.

(9) Make an annual report to the Governor and the Legislature and include such recommendations for changes and amendments in existing law and licensing procedures as are warranted by investigations and research.

(10) Conduct or sponsor a program or programs for research and development of fisheries resources of the State which may include, but not be limited to, biological, chemical, technological, hydrological, processing, marketing, financial, economic, and promotional research and development. The agency may carry out such a program or programs in cooperation with other state agencies, federal, regional, and local governmental entities, or with private institutions or persons.

(11) Establish programs for public education concerning the conser-

vation, utilization, development, and enhancement of fisheries resources.

(b) In discharging the powers and duties enumerated in subsection (a), the [director] [commission] may:

(1) By regulation prohibit, limit, condition, require or establish the use of specified types of fishing gear; the size, number, and quantity of specific fisheries resources that may be taken; the areas to be opened or closed to their taking; the time and manner of their taking; the number of persons or vessels or amount of fishing gear that may participate in harvest of a specified fisheries resource; and may prescribe such other limitations, conditions, requirements, or restrictions as is necessary and appropriate to the policy and purposes of this act.

(2) By regulation authorize the issuance of such licenses, tags, and permits for fishing and may prescribe such tagging and sealing procedures as is necessary to carry out the provisions of the laws and regulations or to obtain information for use in fisheries management.

(3) By regulation prescribe procedures requiring the holder of any license, tag, or permit issued pursuant to the marine fisheries laws to keep records and make reports concerning the time, manner, and place of taking fisheries resources, the quantities taken, and such other information as is necessary for proper enforcement of the laws and regulations or to obtain information for use in fisheries management.

(4) Acquire, introduce, propagate, and stock marine fish species in such manner as will carry out the fisheries resources management programs.

(5) Establish and develop fisheries management areas and prescribe rules governing the use of such areas.

(6) By regulation prescribe fees for licenses, tags, and permits issued pursuant to the law, and user charges for recreational fishing or other recreational uses of lands owned or managed by the agency, unless such fees or user charges are otherwise prescribed by law.

Section 7. [*Procedure for Adoption of Regulations.*]

(a) Any proposed regulation shall be published in a newspaper of general distribution in the area affected by the proposed regulation. Interested persons shall be afforded a period of not less than 45 days after such publication within which to submit written data, views, or comments. Except as provided in subsection (b), the [director] [commission] may, after the expiration of such period and after consideration of all relevant matters presented, promulgate the regulation with such modification, if any, as is appropriate.

(b) On or before the last day of the period fixed for the submission of written data, views, or comments under subsection (a), any person who may be adversely affected by such proposed regulation may file with the [director] [commission] wirtten objections to specific provisions of the proposed regulation, stating the anticipated adverse effect and the grounds for his objections, and requesting a public hearing on such objections.

(c) As soon as practicable after the period for filing objections has expired, if the [director] [commission] determines that the person filing objections and requesting a public hearing may be adversely affected, the [director] [commission] shall hold a public hearing in accordance with subsection (e).

(d) Within 45 days after completion of the hearing, the [director] [commission] shall announce a decision and take appropriate action.

(e) Whenever provisions of this act require a public hearing, the following procedure shall be followed:

(1) The [director] [commission] shall publish in a newspaper of general circulation in the affected area a notice specifying the time and place at which a public hearing shall be held for the purpose of receiving information relevant to the matters identified in the notice of hearing.

(2) The [director] [commission] shall cause a public hearing to be held at the time and place designated in the notice but not less than [ ] days after date of such notice.

(i) The [director] [commission] may appoint a member of the agency to conduct the hearing.

(ii) Any interested person may appear and testify at the hearing.

(iii) The [director] [commission] shall cause a record to be kept of all testimony given at the hearing; said record shall be available for public inspection.

(3) As soon as practicable after completion of the hearing, but no more than [ ] days after, the [director] [commission] shall determine what action must be taken and make his determinations public with a statement of reasons therefor. This record shall be available for the public.

Comment: This section pertains to those States without an Administrative Procedures Act.

Section 8. [*Adoption of Emergency Regulations.*] Notwithstanding Section 7, the [director] [commission] may declare any regulation published under Section 7(a) to be immediately effective, suspended, or otherwise modified, if he determines that due to an emergency situation it is impracticable or contrary to the public interest to defer such actions, and incorporates the determination and a brief statement of the reasons therefor in the publication of the regulation. In such events the opportunity for comment, objection, and a public hearing set forth in Section 7 shall commence with the date of publication of said regulations, and as soon as practicable, but not more than [ ] days thereafter, the [director] [commission] shall publish in a newspaper of general circulation in the affected area his determination as to whether said regulations shall be continued in effect, amended, or terminated, and if amended, the text of said amendment. In the absence of such determination, the emergency regulation shall remain in effect for [ ] after the effective date thereof.

Section 9. [*Intragovernmental and Intergovernmental Cooperation.*]

(a) All state agencies whose activities affect fisheries resources, including, but not limited to, fresh water fisheries, coastal zone management, and water pollution control agencies, shall coordinate their activities which may affect fisheries resources with the [director] [commission] and in turn the [director] [commission] shall coordinate his activities with the aforementioned agencies whenever his activities interrelate with the programs of said agencies. Coordination shall include, but not be limited to, the exchange of information and the filing of copies of any applications, petitions, requests, reports, or other similar documents which may bear upon the responsibilities of any of the aforementioned agencies. The head of each agency is directed to work out details of the exchange of such information and to provide an opportunity for any of the agencies to respond, formally or informally, before final decisions are rendered on matters of applications, petitions, requests, or other similar matters. If any agency has received from the [director] [commission] or the [director] [commission] has received from any agency, comments in writing before a final decision is rendered on the matter of an application, petition, request, or other similar matters, it shall consider those comments in its final decision and they shall be made a part of the record.

(b) In the management of fisheries resources which are shared by the State of [ ] with other States, or with other States and the federal government, the [director] [commission] shall cooperate with the appropriate agencies of said States and the federal government to develop integrated management plans for such shared resources.

Section 10. [*Reciprocity.*] The [director] [commission] is empowered to make the following reciprocal agreements with other jurisdictions:

(1) Rules and regulations with respect to fishing in boundary waters after consultation with the like agency of the neighboring State(s).

(2) A procedure whereby valid fishing licenses issued by the parties to the reciprocal agreements may be used by their licensees within the jurisdiction of either in accordance with the terms of such agreements.

(3) Any individual from a State(s) who has the responsibility of enforcing that State's marine fisheries laws may pursue any person found fishing in the coastal waters of such State in violation of the marine fisheries laws thereof onto adjacent waters of this State and there arrest him and return him to the State where the violation occurred for the purpose of prosecuting him for such violation; *provided,* that such other State shall have enacted legislation giving substantially similar authority to individuals of this State who have responsibility for enforcing the marine fisheries laws of this State relative to persons found fishing in the coastal waters of this State in violation of the marine fishing laws thereof.

Section 11. [*Catch Statistics.*]

(a) The [director] [commission] may by regulation require that every person licensed to take fisheries resources provide information on the species, number, weight, and any other information pertinent to management of the resources taken in a manner prescribed and on forms supplied by the agency.

(b) Every person engaged in the buying, packing, wholesaling, or processing of any fisheries resources within the State shall keep accurate records, books, or accounts showing the species, quantity, and source of fisheries resources.

(c) Every record, book, or account referred to in subsection (b) shall be open for the agency to inspect at reasonable hours.

(d) The agency may audit the records, books, or accounts of any person referred to in subsection (b); and of anyone who takes fisheries resources and ships directly to market in order to determine the quantity of resources taken and other information pertinent to management.

(e) The records obtained by the agency, and the information contained therein, shall except as otherwise provided in this subsection, be confidential, and the records shall not be public records, and, insofar as possible, the information contained in the records shall be compiled or published in such a manner so as not to disclose the business record of any person.

(f) The information in the above records can be provided to other States and regional fisheries agencies provided that those entities have confidentiality provisions that do not disclose the business record of any person.

Section 12. [*Aquaculture.*]

(a) Every person engaged in the business of cultivating fish or plants, whether planting, promoting their growth, or harvesting them, in, on, or from waters and areas of this State, whether public or private, shall be licensed as provided in this section.

(b) A license granting the privilege to cultivate fish or plants shall be issued upon application and the payment of a fee of $ [ ] to the agency. The license shall be valid for a term of [ ] from [ ] or if issued thereafter for the remaining portion of the term.

(c) Any aquaculture in state lands or waters shall be accomplished only under such regulations as the agency may prescribe, and the agency may adopt such regulations as it deems necessary to enforce the provisions of this section.

(d) The agency may prohibit the placing of any strain or species of fish or plants which it considers injurious to the fisheries resources or to the development of the aquaculture industry in the State in any marine waters, including any impoundments, bay, lagoon, or estuary.

(e) The agency may lease to any citizen or domestic corporation of [State] areas in and on state marine and estuary waters, public lands under state marine and estuarine waters, and portions of the intertidal zone upon which to exercise the exclusive privilege of cultivating fish or plants as provided in this section. Use of such areas shall conform to the minimal use of such areas as the agency may establish. No state lands or waters shall be leased unless the agency determines that such lease is in the public interest.

(f) Any citizen or domestic corporation of [State] desiring to acquire the exclusive privilege of cultivating fish or plants in, on, or from state lands or waters shall make written application to the agency for that purpose, designating the particular area desired. The application shall contain the following information:

(1) Location and description of the area by metes and bounds or coordinates, as appropriate.

(2) The species to be managed or cultivated.

(3) A description of the proposed management or cultivation project in sufficient detail to enable the director to determine:

(i) The compatibility of the project with other present or potential uses of the requested areas.

(ii) The degree of exclusive use of the area essential to the proposed project.

(4) That the applicant either owns or has written permission from the owner to use whatever land above the high-water mark and to exercise any riparian right on the underwater lands necessary to successfully carry out the proposed project.

(g) Each application for a lease of state lands or waters shall be accompanied by a fee of $ [ ], which shall not be refundable if a lease is not granted.

(h) The agency, after consulting its records of filings, shall, if the area requested is found not to be leased and the application conforms with this section and upon a finding that the area is on state lands or waters and that a lease would be in the public interest, shall advertise for bids on the lease of the area

(i) No lease shall be for a greater term than [ ] year(s). No lessee shall be granted an exclusive lease to an area that would in the opinion of the agency foster or tend to foster a monopoly. The lease may be renewed after [ ] year(s). The lessee shall have a prior right to meet the best bid and obtain a renewal of his lease.

(j) Ninety days before any water or land is leased, the agency shall cause legal notices inviting bids to be published in a newspaper of general circulation in each county where the water or land or any part thereof is located, describing the area to be leased, the type of operation to be conducted, and inviting bids.

(k) The agency shall award the lease to the highest responsible bidder, if the bid is above the minimum bid established by the agency, which shall not be less than $ [ ] dollars per acre for each fiscal year. The annual rental fee shall be paid to the agency within [ ] days after the beginning of the rental period. If it is not paid within [ ] days after the close of the month in which it is due, a [ ] percent penalty shall be paid. The agency shall declare the lease abandoned unless the lessee can show reasonable cause for failure to pay such rental fees within [ ] days from the beginning of the rental period.

(l) The agency shall promulgate regulations, establishing when a lease is to be deemed abandoned by the lessee due to inactivity, failure to pay fees or taxes provided for in this section, or failure to properly utilize the leasehold.

(m) Upon a lease being declared abandoned, all improvements, buildings, fish, or plants on state lands or waters therein shall become the property of the State by virtue of such abandonment. Until such time as the lands or waters are again leased by the agency, the agency may operate and maintain the area in the best interest of the State.

(n) All leases shall be held subject to the power of the Legislature to increase or otherwise change the fees, taxes, and other charges relating to such lease.

(o) The leasing of state lands or water shall in no way affect public access for recreational purposes to the state lands or waters contained in the leased area, except that access to the area for recreational purposes shall be only in such a manner as to cause no damage to the area or the fish or marine plants cultured therein.

(p) A leased area may be transferred from one person to another person eligible for a lease only upon the receipt of an application by the agency for a transfer, accompanied by the map and fee specified in subsections (f) and (g) and the approval of the agency.

(q) The agency shall by regulation establish requirements for signs or markers to delineate the leased area.

(r) The agency may require such reports from a lessee as it deems necessary to properly evaluate the operations under the lease.

(s) Aquaculture activities, which are authorized pursuant to this section, and such regulations as may be established hereunder, shall be conducted in a manner compatible with other existing lawful uses.

Section 13. [*Ocean Rearing of Anadromous Fish.*]

(a) A person may be granted a permit by the agency under such terms and conditions as the agency may prescribe, to release and recapture domestically reared anadromous fish in state waters.

(b) A public hearing shall be held pursuant to the provisions of Section 7.

(c) No permit will be issued which may interfere with the natural runs of anadromous fish, result in waste or deterioration of fish, or when the proposed operation is located on a stream or river below a state or federal fish hatchery or egg-taking station.

(d) All fish released into the wild under authority of this section, while they are in the wild, will lose their status as private property and may be taken under the authority of a recreational or commercial fishing license.

(e) Any permit granted by the agency pursuant to this article shall contain the following conditions:

(1) Domestically reared anadromous fish released into state waters shall be marked if the agency determines it practicable. Any such marking shall be approved by the agency.

(2) If after a hearing the agency finds that the operation described in the permit and conducted pursuant to this article is not in the best public interest, the agency may alter the conditions of the permit to mitigate such adverse effects, or may cause an orderly termination of the operation under the permit. An orderly termination shall not exceed a three-year period and shall culminate in the revocation of the permit in its entirety. During this period, the permittee may continue to examine and take specified domesticated anadromous fish according to the provisions of the permit, but may not release additional fish.

(3) If the agency finds that the operation has caused deterioration of the natural run of anadromous fish in the waters covered by the permit, it may require the permittee to return the run to the same condition as it was prior to issuance of the permit. If the permittee fails to take appropriate action, the agency may take such action and the permittee shall bear any cost incurred by the agency.

(4) Prior to release into state waters the fish shall be examined by a responsible pathologist approved by the [director] [commission] to determine that they are not diseased or infected with any disease which, in the opinion of the agency, may be detrimental to the fisheries resources.

(5) The permittee shall have the right to divert all fish returning to the stream to an inspection area as authorized by the agency, and shall be allowed to examine all fish for the purpose of identifying domestically reared fish that have returned.

(6) No unmarked fish may be transported from the trapping facility other than to be returned to state waters.

(7) All fish not bearing marks, if required and approved by the agency for the permittee, will be returned unharmed to the stream or river.

Section 14. [*Enforcement, Penalties, and Forfeitures.*]

(a) Any person who willfully commits an act which violates any provision of this act, any regulation promulgated thereunder, or any license or permit issued pursuant to said act or regulation shall, upon conviction, be fined not more than $[ ] or imprisoned for not more than [ ], or both; *provided,* that any person who has committed such violation and who desires to compromise and settle his liability therefor , may appear with any person designated by the [director] [commission] before a court having jurisdiction over the offense and upon terms and conditions acceptable to both and with the approval of the court compromise and settle his liability for said violation; *provided further,* that said compromise and settlement shall not be tantamount to a conviction.

(b) Any person authorized by the [director] [commission] may execute and serve any arrest warrant, search warrant, or other warrant or civil or criminal process issued by any officer or court of competent jurisdiction for enforcement of this act, any regulation promulgated thereunder, or any license or permit issued pursuant to this act or regulation. Such person so authorized may arrest, search, and seize, with or without a warrant, as authorized by law. Any item so seized shall be held by any person designated by the [director] [commission] pending disposition of criminal proceedings, compromise, and settlement, or an action in rem for forfeiture of such item.

(c) All fisheries resources taken in violation of this act, any regulation promulgated thereunder, or any license or permit issued pursuant to this act or regulation, and all fishing gear used directly or indirectly in connection with said taking shall, upon conviction of any person or persons taking said resources, be subject to forfeiture to the State.

(d) The [director] [commission] may revoke or suspend any license or permit issued pursuant to this act or regulation promulgated thereunder and may also, in the case of revocation, deny for a period of [ ] the holder of said license or permit the privilege of obtaining a new or an additional license or permit whenever said holder, within any five year period, has on two occasions violated the following or any combination thereof: this act, any regulation promulgated thereunder, or any license or permit issued pursuant to this act or regulation and further has either been convicted of or has compromised and settled said violations.

Section 15. [*Severability.*] [Insert severability clause.]

Section 16. [*Repeal.*] [Insert repealer clause.]

Section 17. [*Effective Date.*] [Insert effective date.]

The following two appendices are optional sections which a State may wish to adopt in developing or improving their fisheries management program.

Appendix 1

Section [ ]. [*Controlled Entry.*]

(a) Permit required after [month, day, year].

(1) After [month, day, year] no person may operate gear in the commercial taking of specified fisheries resources without a valid entry permit or a valid interim-use permit issued by the agency.

(2) An interim-use or entry permit is not required of a crewman or other person assisting in the operation of a unit of gear engaged in the commercial taking of fisheries resources as long as the holder of the entry permit or the interim-use permit for that particular unit of gear is at all times actively engaged in the operation of the gear.

(3) A person may hold more than one interim-use or entry permit issued or transferred under this act only for the following purposes:

(i) Fishing more than one type of gear.

(ii) Fishing in more than one administrative area.

(iii) Harvesting particular species for which separate interim-use or entry permits are issued.

(b) Terms and conditions of entry permit; annual renewal.

(1) Each entry permit authorizes the permittee to operate a unit of gear within a specified administrative area.

(2) The holder of an entry permit shall have the permit in his possession at all times when engaged in the operation of gear for which it was issued.

(3) Each entry permit is issued for a term of one year and is renewable annually.

(4) Failure to renew an entry permit for a period of two years from the date of last renewal results in a forfeiture of the entry permit to the agency, except as waived by the [director] [commission] for good cause.

(5) An entry permit survives the death of the holder

(6) An entry permit may not be:

(i) Pledged, mortgaged, or encumbered in any way.

(ii) Transferred with any retained right of repossession or foreclosure.

(iii) Attached, distrained, or sold on execution of judgement or under any other process or order of any court.

(iv) Sold to another person without prior approval of the agency.

(c) Fees — The agency shall establish annual fees for the issuance and annual renewal of entry permits or interim-use permits to further the purposes of this section.

(d) Transfer of entry permits — The agency shall adopt regulations and prescribe conditions governing the transfer of entry permits.

(e) Administrative areas — The agency shall establish administrative areas suitable for regulating and controlling entry into commercial fisheries and may modify or change the boundaries of such areas.

(f) Interim-use permit; qualifications — Pending the establishment of the initial number of entry permits under subsection (g) and the issuance of entry permits under subsection (i), the agency may issue interim-use permits under regulations promulgated by it for specified fisheries to applicants who can establish their present ability to participate actively in the fishery for which they are making application.

(g) Initial number of entry permits — The initial number of entry permits for a specified fishery shall be the highest number of units of gear fished in that fishery during any one of the four years preceding [ ].

(h) Optimum number of entry permits — The [director] [commission] shall establish the number of entry permits for each fishery for which it estimates that the optimum number of entry permits will be less than the highest number of units of gear fished in that fishery during any one of the four years immediately preceding [ ], based upon a reasonable balance of the following general standards:

(1) The number of entry permits sufficient to maintain an economically healthy fishery that will result in a reasonable average rate of economic return to the fishermen participating in that fishery, considering time fished and necessary investments in vessels and gear.

(2) The number of entry permits necessary to harvest the allowable commercial take of the fishery resource during all years in an orderly, efficient manner, and consistent with sound fishery management techniques.

(3) The number of entry permits sufficient to avoid serious economic hardship to those currently engaged in the fishery, considering other economic opportunities reasonably available to them.

(i) Issuance of Entry Permits — Following the establishment of the optimum number of units of gear for a particular fishery pursuant to subsection (h), the agency shall adopt regulations establishing qualifications and conditions for issuance of entry permits. The agency shall accept applications for entry permits and issue the permits in accordance with the aforementioned regulations.

(j) Revisions of the number of entry permits.

(1) The agency may increase or decrease the number of entry permits for a fishery when one or more of the following conditions makes a change desirable, considering the purposes of this section.

(i) An established long-term change in the biological condition of the fishery has occurred which substantially alters the number of entry permits permissible applying the standards set out in subsection (h).

(ii) An established long-term change in market conditions has occurred, directly affecting the fishery, which substantially alters the number of entry permits permissible under the standards set out in subsection (h).

(iii) An improvement in the efficiency of fishing methods which substantially alters the number of entry permits necessary to harvest the allowable take of the fisheries resource in an orderly and efficient manner.

(2) If the agency decreases the number of entry permits for a fishery, the number of entry permits may be reduced only under the voluntary buy-back provisions set out in subsection (k).

(k) Buy-Back Program.

(1) When the number of entry permits established pursuant to subsection (h) is less than the number of entry permits outstanding in a fishery, the agency shall establish and administer a buy-back fund for that fishery for the purpose of reducing the number of entry permits to the number established pursuant to subsection (h) within no more than [ ] years, at a rate to be established by the agency.

(2) For each buy-back fund, the agency shall adopt regulations establishing annual assessments on holders of entry permits of not more than [ ] percent of the gross value of the total annual catch attributable to a holder's entry permit, except that the holder of a permit who has made no commercial landings in a given year shall be assessed the average assessed all other holders of the same type of permit in that year. Assessments shall be paid into the specific buy-back fund for which they are collected.

(3) Assessments need not equal annual buy-back fund expenditures within a particular fishery but shall be continued until the buy-back fund for that fishery has been reimbursed.

(4) The agency shall adopt regulations providing for the purchase of entry permits, vessels, and gear at fair market value with money accumulated in the buy-back fund for each fishery. The buy-back program for a fishery shall terminate when the number of entry permits is reduced to the number established pursuant to subsection (h) and the buy-back fund has been reimbursed.

(5) When entry permits and the vessels and gear related to those permits are offered for sale to the agency, the agency may purchase the permits and related vessels and gear at fair market value, if sufficient funds are available in the appropriate buy-back fund.

Appendix 2

Section [ ]. [*Quality Assurance.*]

(a) Duties of the [director] [commission]. The [director] [commission], at his discretion, shall by regulation prescribe minimum standards for establishments and for sanitation and quality control of the harvesting and processing of fish and fishery products. Each set of regulations shall be based upon the particular operational requirements of that species or phase of the industry being regulated and shall conform to all state and federal standards.

(b) The [director] [commission] shall adopt such regulations in accordance with Section 7.

(c) The [director] [commission] shall, in accordance with the most modern public health and food protection practices, establish and maintain effective surveillance and inspection of all segments of the industry for which there is any regulation.

(d) The [director] [commission] shall cooperate with other state and federal department heads or agencies to develop memorandums of understanding detailing duties and obligations of each so that duplication, confusion, and waste will be minimized.

(e) The [director] [commission] or his agent shall have the right to temporarily suspend or revoke the right to process fish or fishery products at any establishment for a period not to exceed 24 hours, whenever it is determined that any provision of this section or any regulation adopted under authority of this section is being or has been violated.

(f) The [director] [commission] or his agent shall have the right to embargo indefinitely any fish or fishery product in any establishment whenever it is determined that the health of the public may be endangered.

(g) Certificates, permits, authorizations. After the effective date of any regulation adopted under authority of this section, no person, firm, or corporation shall process in or for either interstate or intrastate commerce any fish or fishery product in any establishment governed by that regulation unless there is in effect for such establishment a valid certificate, permit, or authorization issued by the [director] [commission].

(h) Application. The application for such certificate, permit, or authorization shall bear such information as may be required by regulation. Any person, firm, or corporation may make application on blanks furnished by the [director] [commission] in accordance with the regulations governing that fish or fishery product which the applicant desires to process.

(i) Issuance of certificates, permits, authorization. The [director] [commission] shall issue any certificate, permit, or authorization if he is satisfied that the applicant has complied with the requirements of the law as set forth in this section, and all regulations adopted under authority of this section. The certificate, permit, or authorization shall bear such information as set forth in the regulations governing the processing of the fish or fishery product for which said certificate, permit, or authorization was issued.

(j) Authorization of holder. The certificate, permit, or authorization entitles the holder to engage in the processing of any fish or fishery product for which that specific certificate, permit, or authorization was issued and at the exact location or address of the establishment listed on the certificate, permit, or authorization.

(k) Expiration of certificates, permits, authorizations. All certificates, permits, or authorizations issued under this section expire at midnight on December 31st of the calendar year in which they were issued, unless sooner revoked or suspended.

(l) Suspension, revocation, denial of certificates, permits, authorizations. The [director] [commission] may suspend, revoke, or deny for any period of any time any certificate, permit, or authorization issued by him, or the right to obtain one, whenever he determines that any of the provisions of law or regulations governing the processing of fish or fishery products and establishments or laws of the State have been violated.

(m) Right of hearing. Any person, firm, or corporation denied a certificate, permit, or authorization under this section may request a hearing with the [director] [commission] to show cause why the particular certificate, permit, or authorization should not be denied.

(n) Any person, firm, or corporation found to be in violation of any of the provisions of this section or any regulation adopted under authority of this section or who continued to violate any portion, shall be requested to appear at a hearing with the [director] [commission] at such time and place as he shall designate, to show cause, if any, why the particular certificate, permit, or authorization should not be suspended.

(1) At the hearing, the person, firm, or corporation may present any relevant facts concerning the alleged violations.

(2) After the hearing, the [director] [commission] may suspend or revoke the particular certificate, permit, or authorization, if he is satisfied that the facts and conclusions justify such action.

(3) Should the person, firm, or corporation fail to appear at the hearing as designated by the [director] [commission], the [director] [commission] may act immediately to suspend or revoke the particular certificate, permit, or authorization, or the right to obtain any.

(o) Right of entry. The [director] [commission] or his agent, whenever fish or fishery products are being processed at any establishment to which a certificate, permit, or authorization has been issued under this section, shall have access to any establishment or part thereof for the purpose of inspection of collection of samples. Denial of access shall be grounds for suspension or revocation of any certificate, permit, or authorization.

(p) Reports. The holder of any certificate, permit, or authorization issued under this section shall make a record of all purchases and sales of fish and fishery products covered by this section and shall file those records with the agency as required by the regulations governing the processing of fish or fishery products.

(q) Products embargoed and condemned. The [director] [commission] or his agent shall indefinitely embargo, condemn, or order to be destroyed any fish or fishery product in any establishment whenever it is determined that the product is of unsound quality, contains any filthy, decomposed, or putrid substance, or may be poisonous or deleterious to health, or otherwise unsafe. The [director] [commission] and his agent shall cooperate with those state and federal agencies having similar responsibility in the protection of public health, in enforcing the order to embargo, condemn, or destroy.

(r) In the event that any fish or fishery product in any establishment is embargoed, condemned, or ordered destroyed, the [director] [commission] or his agent shall, as soon thereafter as practicable, notify the owner in writing of the amount and kind of fish or fishery product embargoed, condemned, or destroyed.

Excerpts from the National Conference on Effective Management of Marine Fisheries

June 24-25, 1975

Role of the States in Fisheries Management

Tom McCall
Former Governor of Oregon

Before beginning I would like to compliment the work of the Task Force and its chairman and vice-chairman. The importance of the Council of State Governments and the National Marine Fisheries Service undertaking an effort such as this cannot be emphasized enough.

As an organized body politic, Oregon has been around only 116 years. But in that time, and especially in the last two decades, quite a bit has changed.

More and more dams have arisen on the Columbia River and its tributaries. Our other coastal streams were closed to salmon gillnetting. Treaties have been interpreted to mean that Indian fishermen on the Columbia are entitled to up to 50 percent of the anadromous fish harvest. Sports fishermen won approval of an initiative prohibiting commercial fishermen from including steelhead in their take, even as an incidental catch. The latter action, which declares the steelhead as a game fish, should make Oregon better equipped to effectively manage coastal fisheries.

Combining Fisheries Agencies

The steelhead is a sea-going rainbow trout — generally in the 2- to 12-pound class, and regarded by sports fishermen as a magnificent fighting fish. I joined in their assessment more than 20 years ago and strived to take the steelhead off the commercial fishermen's list. But that wasn't my sole reason. In the last few years it became apparent that until we settled the steelhead question, we wouldn't be able to merge our Fish Commission, which regulated the anadromous and offshore fishery, with the Game Commission, which regulated sports angling and hunting.

In 1974, the people approved a ballot measure making the steelhead exclusively a game fish. In 1975, a few months ago, the Legislature combined the two agencies into a single Oregon Fish and Wildlife Commission. The merger hasn't formally occurred yet, so I have no experience to relate to you. But the merger makes sense. There are enough problems without having your fisheries agencies fragmented and defensive—or offensive.

One of the key goals of my administration was construction of more fish hatcheries, and we got several. But there was almost a constant bickering over whether a new hatchery would be run by the commercial or the sports side. In addition, research activities were fragmented. There was duplication of effort, too, in environmental protection work and in correction of fish passage problems. Accounting and data collection was always in at least two places. Fisheries education programs weren't integrated.

I'm sure some of us might say that the inland sports fishery—the streams and lakes in the interior—ought to be regulated wholly by an agency oriented toward sports fishing. But a coastal State is hardpressed to make the distinction. Waters from every one of our lakes—except unique Crater Lake—drain into the ocean. We cannot plan solely on a sports fishery in the upper Rogue, for instance, without also considering that some of the salmon produced there will go into the commercial harvest.

So I suggest to you that combining your fisheries agencies, if you haven't already done it, should be strongly and carefully considered before it is rejected.

I'm aware of the suggested Marine Fisheries Management Act proposed by the Council of State Governments for consideration by the coastal and Great Lakes States. The introduction to it suggests the act is concerned with management of fish that are in, or go to, the sea. But the proposal could just as well apply, in many instances, to the entire fishery from mighty ocean to tiny bass pond. We might nit-pick the proposal, but it's essentially sound.

If we can get our sports and commercial fishermen together, they might then attack the real villains, such as the dams on the Northwest's Columbia River. Those dams were a godsend to the Northwest States—States that were industrially deprived. The engineers did pay attention to the insistence of anadromous fish to return home to spawn, and in most cases fish ladders were constructed.

But the downstream migrant was left to shift for itself. As a result, fewer and fewer migrants actually survive the trip to the sea. They're tossed and tumbled; they even get the "bends" from nitrogen supersaturation. The dams now have been labeled as most likely to succeed in exterminating Columbia River salmon and steelhead runs.

A researcher working for the Pacific Northwest Regional Economic Development Commission says it would cost $8 million per dam to correct the problem. But he also described this as "a relatively small investment considering the value of the resource."

So you of the coastal States with rivers dammed or about to be dammed should insist that the onshore-offshore fishery—both a food and fun resource—be protected for all time.

Intergovernmental Cooperation

Well, let's assume you have a strong single agency, or at least some method of coordinating state fisheries activities. Your next worry is the activity of the State or nation next door. The Council of State Governments' suggested act speaks to that, too. It says, in effect, that the state fisheries agency should cooperate with other States and the federal government to develop integrated management plans for shared resources.

This is marvelous language, and it would be decent if all the States and nations lived up to the admonition. So far as I've been able to see, however, fisheries management is inclined to be more parochial than effective.

The State of Washington declared the steelhead a game fish before Oregon, and both Washington and Oregon kept Idaho from becoming a member of the regulatory Columbia River Fisheries Compact until recently—even though Idaho produced a majority or near-majority of the salmon and steelhead fed into the Columbia system.

California fishermen keep insisting on harvesting salmon that Oregon sent to sea, even though the salmon are immature. The last time I dealt with California on that subject, I said our fisheries biologists were going to see to it that Oregon salmon would never again darken California's door. We were, I said, going to develop a "right-turning" salmon, so that on leaving the Columbia these fish would go north toward Washington and Alaska, make their swoop somewhere around the 45th parallel (which is well above the California border), and then come home where they belong.

We didn't develop the right-turning salmon, though, and the problem hasn't gone away. And as more and more pressure is brought to bring home the bacon from the sea, interstate and international rivalries will heat up.

I have proposed that this country adopt, as a minimum, a "three-fish" policy:

Fish on the continental shelf that are year-around inhabitants belong to the bordering State;

Anadromous fish, such as salmon, belong to the State of their origin; and

Pelagic species, such as tuna, are fair game to all States and nations.

By saying "fish" I also mean food fishes such as lobster and shrimp and, hopefully, the policy I've suggested would square with the Massachusetts demand for a 200-mile lobster protectorate. But probably not. When we get into numbers—a 3-mile, 12-mile, 50-mile, or 200-mile limit—or into politics—meaning the limit is how much of the sea you can forcefully protect—we're not being responsible. We have to define our problem within a world context, not just within the context of an individual State.

I've said that a minimal policy of the country should be to let anybody take tuna, anytime, anywhere. That policy has to have its "ifs," "ands," or "buts." Since man is smarter than tuna fish, he also is able to make the species extinct. Since man is smarter than the spiny fin lobsters that roam around the Bahamas, he also can make that species extinct.

The June issue of the *National Geographic* declares that lobster brings $15 million a year to Florida and Bahamas fishermen. It also says: "Increasing numbers of fishermen, however, are threatening the state's lobster population; in the Bahamas the crop is already declining. Florida and the Bahamas impose strict regulations, but no one controls lobstering in international waters." The State Department has worked to establish, by agreement, fisheries conservation zones. They seem to be fairly well, maybe even perfectly,

Commercial salmon troller off the coast of Oregon.

honored by foreign fishing fleets. This is hardly enough. The principal foreign fleets off the Oregon shore—the Soviet Union, Poland, and South Korea — also seem to honor our request not to take salmon. They're taking largely a fish not wanted by our own fishermen, the hake. But they also scoop up bottom fish that we *do* want.

Most of the ocean already is a desert; now the threat is that the continental shelf will become a desert as well. Two rounds of the Law of the Sea Conference have been held. The nations have largely played for position — the developed nations reject any proposal that might hinder their peaceful passage through narrow straits; the developing nations seem to want an International Seabed Authority.

Somehow, it all has to work out in a rational way. There are so many people on this globe that we now must do something few of us old duffers ever thought we'd have to do: manage the seas.

We used to have vast frontiers and a job for everyone willing. We had "unlimited" natural resources waiting to be tapped. Great Britain was, and then America became, the lord of the seas. But our lordship was largely for defensive purposes. Now that other means of achieving security, or making war, exist, we are no longer the lord of the sea—not of the sea that has food in it or beauty in it.

This Nation ignores the protein that is in the hake the Soviet Union scoops up off our shore. This Nation has, however, quit the harvest of whales for fear of their extinction, and for love of their beauty—but we also haven't made the whalers of other nations desist from their depredation.

Conclusion

It may seem that I have gone astray from the conference theme of "Effective State Management of Coastal Fisheries." But I haven't. Effective state management *does* start with the State itself, a single agency or a coordination by executive fiat of fisheries agencies. But that state management of fisheries cannot be effective without correlation of fisheries with water quality controls, dam construction, and land use planning. It also cannot be effective if our Federal government does not take prompt, stern steps to protect the renewable resource known as fish. Finally, it cannot be effective unless we demonstrate — by facts, not by argument — that the foreign nations are depleting for their own purposes a resource that is limited in supply and that all the world should be able to depend upon for a portion of their protein supply.

Role of Compacts in Fisheries Management

Panel Session

Atlantic States Marine Fisheries Compact

From 1937-41, a series of conservation conferences led to the eventual formation of the Atlantic States Marine Fisheries Compact. Some of the participants at these conferences were the Council of State Governments, administrative and legislative delegates from the Atlantic States, and the U.S. Bureau of Fisheries which later became the U.S. Fish and Wildlife Service.

The stated purpose of this compact is to "promote the better utilization of the fisheries, marine, shell and anadromous, of the Atlantic seaboard by the development of a joint program for the promotion and protection of such fisheries, and by prevention of the physical waste of the fisheries from any cause. It is not the purpose of this compact to authorize the States joining herein to limit the production of fish or fish products for the purpose of establishing or fixing the price thereof, or creating and perpetuating monopoly."

In the original draft of the compact it was recommended that the joint commission act as a regulatory body. However, at the second conservation conference, it was recommended and adopted that the joint commission only be a body which could make recommendations and would have no regulatory authority. The primary reason for this was that the conference participants felt that it would not be politically acceptable for an interstate compact to have this type of authority.

The third conference in this series of conservation conferences resulted in the recommendation that an advisory committee be formed to advise the commissioners on the ASMFC. The primary benefit of this advisory committee is that it has enlarged the overall participation in the ASMFC.

Amendment 1 and the ASMFC

Subsequent to the original formation of the ASMFC, it was found that despite possible political opposition there was a distinct need for the commission to have some type of regulatory authority. To respond to this need Amendment 1 was proposed and adopted by nine of the 15 member States to give the commission some regulatory authority. For the greatest part of its existence, the amendment has been considered to be of a highly innocuous nature. However, in 1972 under the authority of this amendment the North Shrimp section was formed to regulate this particular fishery within the Gulf of Maine. The States which are involved in this are New Hampshire, Massachusetts, and Maine.

Left: Oyster dredging off the North Carolina coast.
Bottom: Digging for clams.

The ASMFC Today

Most individuals involved in the ASMFC perceive it to be a practical-type commission, rather than a super-governmental agency. The primary reason for this is that the commission utilizes existing state/federal agencies in a joint effort to solve those problems confronting this particular region. Despite the existence of this organizational vehicle to discuss and resolve many of the mutual problems confronting the Atlantic States, there has not been a great deal of progress. To illustrate this point all that one has to do is consider many of the problems that were considered in the 1937-41 meetings, e.g., lobster management, uniform lobster size, and allocation between commercial and recreational. Despite the lack of comprehensive progress in many of these regional problems there is still a consensus that the tools for carrying out a regional management program have been developed.

The Gulf States Marine Fisheries Commission

Up until now the GSMFC has been remiss in not informing the various State Legislatures on the various problems confronting fisheries interests. However, the commission has begun to reverse this trend and has started to educate the Legislatures about the importance of wise management and utilization of fisheries resources. One of the primary reasons for this new emphasis on educating the Legislatures is the fear that if the States do not take some constructive steps toward improving fisheries management, the federal government will intervene in the management of this crucial issue.

The Eastland Resolution

Under the auspices of this resolution, the GSMFC plans on doing the following: (1) examining what authority various agencies have which deal directly or indirectly with marine fisheries; and (2) determining whether or not agencies are using their statutory authority, and if they are not using their authority, why this is happening. It is hoped that by using this resolution the commission can direct fisheries management to where it will be productive.

Underwater Obstruction Committee

This committee has been formed in order to coordinate state and industry action in attempting to solve problems in this area. With the assistance of industry representatives, the National Marine Fisheries Service, the Bureau of Land Management, and other groups, this committee helped establish new regulations for the outer continental shelf. It appears to be a general consensus among members of the Gulf States Marine Fisheries Commission that this type of committee should be considered by the other commissions.

The Pacific Marine Fisheries Commission

Primarily, there have been four objectives of the PMFC; in three out of four the commission has been successful.

The commission has been successful in providing energetic leadership in recognizing and resolving fisheries problems. Examples of this are: (1) the commission has been active in assisting States review and make an input into the National Marine Fisheries Plan; (2) with the possibility of extended jurisdiction it has explored alternative management schemes. Another area of success for the commission has been in the coordination of research and management projects. An example of this is the Dungeness crab project which is under way. The third success of this commission has been its capability to provide a forum for the member States to present their position on various issues which are communicated to both the state and federal governments.

The area in which the commission's activities has not been as successful has been in the area of proposing compatible fisheries regulations based upon scientific evidence and with full consideration of biological, ecological, recreational, esthetic, social, economic, and political aspects. There have been two major reasons why the commission has been great at proposing but not at accomplishing in this area. First, there has been an absence of effective enforcement mechanisms. The reciprocal enforcement idea which is presented in the Marine Fisheries Management Act may be an effective means to overcome this barrier. Secondly, as long as management decisions are made in the political arena, with the special interest pressures that are exerted in this area, rather than by people who are charged with management responsibility, there will be some blockages.

Commission Organization

Presently, there are 15 commissioners, a scientific staff, and advisors from industry. There are standing committees built around fisheries which are of the greatest importance to the region.

The advisory committees are organized on a state basis, with advisors selected to represent various user groups. These advisors are appointed by the state fisheries agency, the Governor, or the two working together. It is noteworthy that these advisors have their expenses paid by the commission; this eliminates the necessity to consider the financial well-being of a prospective advisor. By taking this tack the commission assures itself of having a broad representation of user groups among its advisors.

A second characteristic of the PMFC, which makes it unique, is the fact that the commission pays the expenses for all commission business, including attendance at annual meetings for three commissioners, seven advisors, and five scientists from each State. Therefore, the commission insures itself a good attendance at its annual meetings. Also, this enhances the potential of having individuals who are chosen for their representative capability rather than for the size of their savings account.

Another unique quality of the PMFC is the fact that the executive director has the option to call special meetings of scientists when they are needed for a particular problem. The only constraint in this is that the State has to approve the leave of absence for the scientist.

The funding of the PFMC is based on an annual budget which is approved by the executive committee. Once this occurs the amount approved is prorated according to a ratio among the States.

Resolutions which the PMFC considers are generated by the various advisors and the States. It is a rule that proposed resolutions be circulated 30 days before a meeting. The commission will pay the in-state expenses of commissioners, advisors, and scientists so that they can confer with various groups within the State to determine whether or not there are special areas of concern in the proposed resolution. The use of resolutions has proven to be highly popular with user groups because they are given the opportunity to have the first input. It is at the annual meeting where these proposed resolutions are debated and voted upon. Once a resolution has been finalized, the commissioners do not have to accept the advice which is contained within the resolution.

Operations of the PMFC

In discussing the operations of the PMFC, it is essential to be cognizant of the limitations under which this and other compacts were established. Essentially, they represent States for the purposes of communication and planning and not as enforcement agencies, with the possible exception of Amendment 1.

With the PMFC there are three operational functions which it perceives itself as serving. First, it serves as a means of vertical and horizontal communication among the States, the federal government, local authorities, and the various user groups. Second, it serves as an entity for the planning and management of research efforts. An example of this could be a project in a particular area, such as the state-federal fisheries management program. Finally, as far as the States concur, the commission can assist in the implementation of various programs. Usually, this does not go into the enforcement field. However, it does often relate to a particular project, for example, the management and funding of a particular research effort.

The Great Lakes Fisheries Commission

The management of fisheries in the Great Lakes is similar to the management of marine fisheries resources, due to the size of the lakes. It appears as though problems, as well as solutions, arise more rapidly in this area than with their marine counterparts.

The Great Lakes Fisheries Commission is a treaty organization. There are eight commissioners, four from the United States and four from Canada.

Each national section has one vote and two are required for action. There is a full-time secretariat to provide service for the commission. The express duties of the commission are to: (1) coordinate research for the optimum sustained yield of fisheries within the Great Lakes, and (2) directly control sea lampreys. The primary weakness of this commission has been the fact that it has a technical makeup; therefore, there has been an absence of strong political support to solve the problems of eight States. It should be noted that there is also a problem of coordination in the U.S. waters of the Great Lakes due to the fact that there are eight States and two major federal agencies involved, while the Canadian counterpart only has to deal with one province and one federal agency. Third, the affected States are not members of the commission. Fourth, they do not have the travel funds, such as PMFC; therefore, it makes it difficult to handle emergency situations or formal management actions between the various States.

At this time, the eight States hope to arrange a caucus to represent the various positions of these States. It is hoped that in taking this action the needs of this area will be more forcefully presented.

The State-Federal Fisheries Management Program

Richard H. Schaefer
Chief
Fisheries Management Division
National Marine Fisheries Service
National Oceanic and Atmospheric Administration
U.S. Department of Commerce

The State-Federal Fisheries Management Program, as now conducted, is an evolving experiment in the cooperative management of interjurisdictional fisheries. In the context of the program, fisheries management can be defined as the establishment, administration, and enforcement of regulatory regimes, developed and maintained through the acquisition, analysis, and application of relevant resource and fisheries data, which will provide for the conservation, rehabilitation, and rational utilization of fisheries resources. The program was created to prime and catalyze the development and implementation of effective management plans for fisheries resources over their entire geographical ranges of distribution. Therefore, close intergovernmental cooperation between and among the States and the federal government in management planning, regulation, and enforcement regarding commonly shared fisheries resources is a program requisite.

In order to better understand the criticality of the relationship between the legislative proposal being presented here today and the State-Federal Fisheries Management Program, it is necessary to briefly review the program's short history. The program was formally established in 1971. It was born, and continues to evolve, out of a recommendation made in 1969 by the President's Commission on Marine Science, Engineering and Research, otherwise known as the Stratton Commission. Charged with undertaking an intensive investigation into a broad array of marine problems the commission observed, with regard to our Nation's fisheries, far too many cases of outmoded fleets, excessive harvesting capacities, overexploited stocks, declining catches, user conflicts, high unemployment, low incomes, etc. It concluded that these conditions could be systematically traced to two basic causes:

(1) That limited fisheries resources are considered common property, available to unlimited access by users.

(2) That fisheries are regulated (or not regulated) under split or multiple jurisdictions, with no single focus of management responsibility.

Among other things, the commission suggested that the rehabilitation of domestic fisheries depended on the elimination of overlapping, and oftentimes conflicting, laws and regulations which have tended to impede even those fisheries which were economically viable. It recommended "a definitive review and restructuring of fisheries laws and regulations, and the creation of a new framework based on national objectives for fisheries development and on the best scientific information." Thus, the State-Federal Fisheries Management Program.

Principal Objectives

The goal of the State-Federal Fisheries Management Program is to effect the rational management of domestic interjurisdictional fisheries through the development and implementation of comprehensive fisheries management plans so as to optimize social, recreational, and economic benefits on a sustainable basis. The principal objectives to attain this goal are:

(1) To develop and maintain an institutional structure that facilitates cooperative state-federal management planning and action, with advice from resource users.

(2) To develop and promote appropriate legislation that provides the necessary regulatory authority to effectively manage fisheries.

(3) To design and implement appropriate program policies and planning guidelines that provide for shared decision-making and positive, timely management action.

In essence, program goals and objectives will have been achieved when there is optimum utilization of biological stocks on a sustained basis; when commercial fisheries are economically viable; when recreational angling opportunities have been enhanced; and when the American consumer is provided with a continuous supply of high-quality seafood products.

To date, much has been accomplished and much experience has been gained through program operations. Yet, we have barely scratched the surface of the problems associated with achieving effective management. Let me just briefly describe to you what has been done to date.

Developing and Maintaining an Institutional Structure

With regard to the first program objective, i.e., developing and maintaining an institutional structure, state-federal fisheries management councils have been established in each of the five national marine fisheries service regions distributed around the coasts of the United States. Consisting of key fisheries administrators from coastal States in each region, and the regional director of the National Marine Fisheries Service, these councils function as joint state-federal recommendatory and decision-making bodies at a policy operational level. As currently practiced, it is the responsibility of these councils to identify appropriate fisheries resource "targets" for joint state-federal management; to jointly effectuate the development and implementation of comprehensive management plans for those "target"

A catch of menhaden.

resources; to establish appropriate working committees of planners, biologists, economists, social scientists, etc., to develop the management plans under council guidance; and to adopt and implement such plans, to the extent possible, including the promulgation of appropriate regulations and their enforcement through the authorities of the individual States represented on the councils. Since mid-1972, when the American lobster was identified as the first state-federal fisheries management "target," nearly $1.5 million of federal program funds have been provided to support the operations and activities of regional councils around the Nation, including expenditures for planning, travel, and data acquisition.

In addition to the American lobster, six other fisheries resources or resource "complexes" have also been designated as state-federal management "targets." These are the northern shrimp in the Gulf of Maine, the surf clam of the mid-Atlantic coast, penaeid shrimp in the South Atlantic States, menhaden in the Gulf of Mexico, certain coastal fishes of southern California, and the Dungeness crab on the Pacific Coast. In addition, Alaskan king and tanner crab, Pacific Coast shrimp, and Pacific troll salmon are also under serious consideration as near-future state-federal management "targets." Phases of management plan development and implementation vary considerably at this time from fishery to fishery, i.e., from merely "target" identification in a few, to completion of plan development and partial implementation of regulations in others. It was recognized very early in the development of the State-Federal Fisheries Management Program that, to achieve its major goal, more effective regulatory authority and mechanisms must be developed and promulgated at both the state and federal levels. Continued lack of effective management control, both in many States and at the federal level, has proven to be a major program weakness.

The problems of resource depletion, economic problems of fishermen, and conflict between users are of great concern to you and your constituents. We also seek resolution of these problems. We share with you the broad goal of conserving these valuable fisheries resources which generate benefits for commercial fishermen, recreational fishermen, and local communities. Further actions are necessary by the States and by the federal government to strengthen this cooperative fisheries management system and to make it more effective.

Developing and Promoting Appropriate Legislation

What is perceived as needed at the federal level is general management authority over fisheries seaward of the territorial seas, and accessory legislation that would provide a funding base and specify other institutional, operational, and planning requirements for the State-Federal Fisheries Management Program. Among other things, general management authority would enable the federal government to establish regulations in areas outside state jurisdiction that could complement agreed-upon and recommended regional council policies for interjurisdictional fisheries resources. The accessory legislation would provide specific statutory endorsement for the state-federal management concept, would solidify state-federal institutional arrangements, and would improve management plan development and implementation mechanisms.

In 1973, the Administration-supported H.R. 4760 and S. 1069, the "High Seas Fisheries Conservation Act," was introduced into both houses of the U.S. Congress. This legislation would have provided federal management authority over fisheries in the contiguous fisheries zone and over U.S. vessels on the high seas. For a variety of reasons, however, this legislation was not enacted.

The legislative situation at the federal level, however, as it relates to fisheries, may soon change. There are several legislative proposals presently before the U.S. Congress that would extend the U.S. fisheries zone seaward from 12 miles to 200 miles. Several of these proposals, in one way or another, provide for some form of federal management authority in the zone of extended fisheries jurisdiction.

It is entirely conceivable that should one of these proposals be enacted and signed into law, not only would it provide long-sought federal regulatory authority, but it would also provide the specific legislative base sought for the State-Federal Fisheries Management Program.

At the present time, however, the implementation of management plans and uniform or coordinated regulations over common fisheries resources remains almost exclusively dependent on the regulatory authorities of the participating States. To date, this arrangement has not been very satisfactory. Most state fisheries administrators involved in program activities would concur that while it has been possible for them to reach general agreement in Council deliberations on many basic policies, standards, and regulations for "target" fisheries, it has been difficult to obtain the timely approval of the state legislative authorities. This is proving to be a significant obstacle to the achievement of program objectives.

Coastal States were granted management authority over fisheries resources in the territorial seas under the Submerged Lands Act of 1953. The way in which such authority is exercised is extremely crucial to our Nation's fisheries resources. Indeed, nearly 60 percent of all U.S. commercial fisheries landings are harvested from within the territorial seas and, further, about 80 percent of all U.S. commercial landings are taken from within 12 miles of the coastline. While some of those fisheries resources are strictly intrastate, a majority are of an interjurisdictional nature. Therefore, the actions or inactions of any single State over such migratory or broadly distributed resources can be far-reaching.

Designing and Implementing Program Policies

The methodology and organizational arrangements for executing state fisheries management authority vary considerably from nearly complete legislative control in some States, to nearly complete delegation of authority to the management agencies in others.

In general, three distinct mechanisms for making decisions and implementing fisheries management actions can be identified among the coastal and Great Lakes States. Further, for reasons not fully understood, there seem to be distinct regional differences in state fisheries management systems. For example, the States bordering on the Gulf of Mexico and on the Pacific Ocean exhibit a strong preference for delegated fisheries management responsibilities. Although, as one would expect, there are variations, it is interesting to note that eight of the 10 States in those regions have chosen to delegate these responsibilities to a board, a commission, or even to the fisheries agency director. In comparison, only four of the 14 Atlantic Coastal States have opted for a comprehensive delegation of management authority. The remainder of the Atlantic Coastal States have elected to either delegate no regulatory responsibility, or only some responsibility which is most often subject to legislative review or, in the extreme case, legislative approval.

Moreover, among those States which have elected to retain strong legislative control over fisheries matters, several conduct only very brief annual legislative sessions, while in others the Legislatures only convene biennially. The point of this very brief analysis is that the dynamics of fisheries and fisheries resources demand quick and flexible management responsiveness to rapidly changing situations. Good fisheries management requires a capability, for example, to open and close fishing seasons or zones; modify size, sex, and catch limits; and respond to other crucial situations in very short periods of time. Indeed, the continued well-being of certain fisheries and fisheries resources depends upon such a capability. States whose fisheries management agencies have largely been delegated such regulatory flexibility are in a much better position to deal effectively and decisively with fisheries problems as they develop. States that must rely exclusively on the legislative process to respond to the day-to-day "housekeeping" problems facing fisheries managers may find that during the time required to react, the problems have progressed beyond the bounds of reasonable solution. Therefore, insofar as the State-Federal Fisheries Management Program is concerned, there is a need at the state level to focus largely on improving the management-response time and regulatory flexibility of the fisheries agencies.

Conclusion

In addition to that basic reform, there are other specific needs that also require attention. There is a need to:

. . . More clearly define the legal relationship between the State, in its sovereign capacity, and the fisheries resources.

. . . Broaden the objectives of fisheries management to include economic and social, as well as biological objectives.

. . . Mandate, rather than simply permit, close intergovernmental cooperation between and among States and the federal government.
. . . Facilitate adequate licensing and catch reporting requirements from resource users.
. . . Encourage advisory input from resource users, conservationists, and other interested private citizens in the decision-making and regulatory processes.
. . . Establish effective and, where appropriate, reciprocal penalty and enforcement deterrents.

The legislative proposal before you today, as developed by the Council of State Governments' Task Force on Effective State Marine Fisheries Management Programs, contains and recommends such provisions.

The need for such improvements in state legislative and regulatory mechanisms is greater now than ever before. The legislation that would provide management authority for the federal government in a zone of extended fisheries jurisdiction would probably also provide a legislative basis for "hardening" of the State-Federal Fisheries Management Program. For example, if current thinking prevails, some form of regional fisheries management councils, composed in major part by state representatives, would probably be formalized to supplant or be reconstituted from the existing regional councils currently established under the State-Federal Fisheries Management Program. Furthermore, pressure from the U.S. citizenry and a mandate from the U.S. Congress would probably require rapid acceleration of the establishment of integrated state-federal management regimes for interjurisdictional fisheries. Such timely implementation of integrated, resourcewide, management regimes can only be accomplished if regulatory mechanisms are streamlined at both the state and federal levels.

With regard to the proposed legislation I would like, in conclusion, to applaud the work of the Task Force on Effective Fisheries Management Programs, the Council of State Governments, and the project staff. Their dedicated efforts over the past year have, indeed, been outstanding and very much appreciated. The product represents the best thinking of some of the leading state fisheries administrators and legislators from around the coastal areas of the United States, including the Great Lakes. Whether adopted in whole or in major part, its recommended provisions and mechanisms would do much to improve intrastate, interstate, and state-federal fisheries management effectiveness. The relationship between such legislation and the future well-being of many of our coastal fisheries and fisheries resources seems quite clear. Certainly the attainment of the goals and objectives of the State-Federal Fisheries Management Program depends, in major part, on it. Speaking for the program, I hope that you will look upon the recommended legislation with as much enthusiasm, and lend it as much support, as we think it merits.

Limited Entry—The Alaska Experience

Roy Rickey
Chairman
Limited Entry Commission
Alaska

The main impetus behind the enactment of limited entry legislation was the problems confronting the salmon fisheries in the State. The primary problem was a radical increase in the amount of gear being used to fish a resource which is now only slowly replenishing itself with the assistance of the State. Secondly, it was strongly believed that if the State did not take some form of affirmative action, the federal government would intervene. The latter possibility was extremely unattractive to the State and the fishermen involved.

In 1972 the Alaska Legislature appropriated funds to be used by a four-person study group with several advisors to: (1) suggest some type of legislation which would address the problem; and (2) do extensive research into the economics of the fishery. A result of this latter effort was a publication containing those statistics which are salient to the development of a limited entry program.

In the 1973 legislative session, the Governor introduced a bill which was the work of this study group. Several versions later, the present Alaska legislation was enacted. Presently, the law has withstood its first legal challenge in a superior court. This favorable ruling has been appealed by the plaintiffs and will be heard by the Alaska Supreme Court in the fall of 1975. If the State Supreme Court's ruling is unfavorable, the State intends to pursue the issue up to the Supreme Court of the United States.

How the Act Works

There are two areas which make this act most unique. First, the commission sets the fees for the permits, which reflect the cost of administering this program. Secondly, the general taxpayer is not paying for this program. The fishermen who are involved in the program are the source of funding. They are paying this entry permit fee in addition to their licensing fee.

The act applies to only 19 fisheries. Out of that number only eight have been designated distressed fisheries. When a fishery is designated a distressed fishery, a reduction plan goes into effect. This means that the State is attempting to reduce the amount of gear used to the optimum level for that particular fishery. The mechanism that is used in the reduction plan is a voluntary buy-back system. In this system, the State enters into the market and competes with other individuals for the purchase of a person's permit.

This system is funded by the fishermen in a particular fishery. For each buy-back fund, the commission may assess holders of entry permits for that particular fishery at a rate of not more than 7 percent of the gross value of the total annual catch attributable to a holder's entry permit. This particular feature could be a scary thing for the fishermen. Presently, the program is not at a buy-back stage; however, in six months they will be hiring a buy-back manager.

With the remaining designated fisheries, the commissioner has estimated that the optimum and maximum number of entry permits are synonymous. Therefore, there is no need to institute a buy-back system.

These designations can be changed by the commission. However, this can only be done on a long-term basis and for primarily two reasons. The optimum number of entry permits can be increased or decreased for a particular fishery when one or more of the following conditions make a change desirable, considering the purposes of the act.

(1) When an established long-term change in the biological condition of the fishery has occurred which substantially alters the optimum number of permits permissible, applying the standards set out in Section 16.43.290.

(2) When an established long-term change in market conditions has occurred directly affecting the fishery which substantially alters the optimum number of entry permits permissible.

At this time there has been no definition as to what "long-term" means.

Sec. 16.43.290. Optimum number of entry permits. Following the issuance of entry permits under §270 of this chapter, the commission shall establish the optimum number of entry permits for each fishery based upon a reasonable balance of the following general standards:

(1) the number of entry permits sufficient to maintain an economically healthy fishery that will result in a reasonable average rate of economic return to the fishermen participating in that fishery, considering time fished and necessary investments in vessels and gear;

(2) the number of entry permits necessary to harvest the allowable commercial take of the fishery resource during all years in an orderly, efficient manner, and consistent with sound fishery management techniques;

(3) the number of entry permits sufficient to avoid serious economic hardship to those currently engaged in the fishery, considering other economic opportunities reasonably available to them. (§ 1 ch 79 SLA 1973)

In determining who would be eligible to obtain an entry permit, the State developed a point system. There are presently two ways in which to obtain a license. First, if an individual fished prior to January 1, 1973, and owned his vessel and gear, he would receive 21 points and be eligible to apply for a permit. This retroactivity clause has been a primary cause of many complaints; however, the Legislature did this in order to avoid a flood of gear being used in a particular fishery prior to the enactment of this bill. Second, an individual is eligible to apply for an entry permit if he is financially 100 percent dependent on that particular fishery.

At the outset the philosophy behind this act has been that everyone would be treated the same. The primary reason for this is that the administrators are aware of the important constitutional issues that are involved with a program of this nature. Therefore, the administration of this program has been extremely delicate.

Under this program the entry permits are considered personal property. Therefore, according to the act:

> (g) An entry permit may not be:
> (1) pledged, mortgaged, or encumbered in any way;
> (2) transferred with any retained right of repossession or foreclosure; or
> (3) attached, distrained, or sold on execution of judgment or under any other process or order of any court.

This section dealing with transferability was a big stumbling block for the Legislature. The primary reason for this is that the U.S. Bankruptcy Act may supersede this particular clause. The rationale of inserting the clause dealing with pledging, mortgaging, or encumbering was that the Legislature wanted to prevent the processing companies from gaining control of the permits. In Alaska's salmon fisheries, many of the vessels are not owned by the fishermen but rather by the companies.

Promulgation of Regulations

During this phase the commission held 60 meetings with fishermen. Most of these meetings were public hearings which helped establish the basis for the regulations. When the regulations were proposed, the commission once again consulted with the fishermen about the fairness of the proposed regulations. Based on these meetings there were some changes made in the final regulations.

The Commercial Fisheries Commission

There are three commissioners appointed by the Governor to four-year, staggered terms. These individuals must be confirmed by the Legislature. Presently, 22 out of 24 staff positions are filled. There is a research staff, a permit-issuing section, an economist, a fish and game coordinator, three attorneys, and one person designated by the Attorney General's office as an advisor. Therefore, in legal matters the commission is well equipped to keep a finger on where it is legally.

The commission was established as a distinct entity at the outset in order to avoid disruption. However, it may be incorporated into the Department of Fish and Game at some later date.

A Russian fishing vessel off the North Carolina coast.

Statistics: Who Counts What and Why

John P. Harville
Executive Director
Pacific Marine Fisheries Commission

In February 1973, representatives of some 50 nations and 11 international commissions met in Vancouver, British Columbia, for the Food and Agricultural Organization's (FAO) Technical Conference on Fisheries Management and Development. The last comparable meeting was in 1955 in Rome, and it considered only conservation, not developmental aspects. During the intervening 18 years, world annual production of fish and shellfish has grown from 28 million tons to approximately 70 million tons. The major fishing nations have developed distant-water fishing fleets of a capacity and technological sophistication undreamed of a few decades ago. These highly mobile, floating catcher and factory ships have the unquestioned capacity to harvest major target species to the point of no return as an economic resource. Indeed, the capacity to harvest 100 million tons per year from the world's oceans is clearly with us today, and this is an upper limit set by many scientists for sustained yield from conventional species.

Concurrent with this burgeoning capacity for overexploitation by man, the oceans and their tributary systems have diminished in their capability to produce the harvest upon which man depends. Diking, filling, and other development of marshes and wetlands which form the production base of estuarine ecosystems; massive pollution of rivers, bays, and the ocean itself; competitive uses for other purposes of choice spawning and rearing environments—these and other outgrowths of man's progressive encroachments on the sea have reduced the ability of that great ecosystem to recycle the raw materials and replenish the resources upon which our fisheries depend.

For these and many other related reasons, this landmark 1973 FAO International Conference was convened under terms of reference to "consider scientific and technical principles and methodology for both fishery management and fishery development." After two weeks of concentrated debate and discussion the conference approved 11 recommendations, of which the first three are particularly pertinent to this conference:

1. *Planning.* The conference recommended an integrated approach to fisheries development planning, and emphasized that it was essential to take account not only of scientific data and information but also relevant social and economic factors.

2. *Fisheries Data.* The conference stressed that rational fisheries management and development require adequate and timely statistics and other data relating to fisheries. Therefore, it recommended that governments concerned insure that national and international management institutions are

provided with adequate data not only on the resources of the catch but also on the fishing effort. It further urged that governments and international fishery bodies take steps to improve fishery data systems, and to adopt common standards.

3. *Need for Early Management Action.* The conference noted that management to protect resources and maintain the yield is already overdue in many instances and should indeed always be started early in the development of a fishery.

For all of us with responsibility for fisheries research, management, and development, this international conference established a number of important guidelines. First, in the context of integrated management planning, it was the clear consensus of our international scientific leaders that fisheries management can no longer remain preoccupied only with biological aspects of protecting the resource. Fisheries management for the latter quarter of the 20th century must be concerned with the *people* side of the fisheries definition as well as the *fish* side, and must accept, as concurrent objectives with resource protection the enhancement of economic yield from the resource and the satisfaction of broad socioeconomic goals of the people they serve.

With respect to fisheries data needs, the conference emphasized the importance of an adequate data base for management decision-making. Participants were critical of existing major gaps in this data base, but they were even more critical of the lack of compatibility and capacity for data-sharing in data produced by our agencies and institutions. The conference urged that steps be taken immediately to remedy this situation. Further, it is clear that this data is needed on a global, not parochial scale.

Closely related to this need for a more comprehensive and compatible data base was insistence on prompt management action. This has several connotations for practical fisheries research and management. First, it tells us to equip our data systems with the technology of the 1970s, to utilize the power and the speed of the computer in order to remove the often unconscionable lag-time between collection of data and its presentation for analysis. Second, it suggests that data analyses should be presented in a manner that will lead to management decision-making—that the systems analysis approach of the engineer be applied to organization of data into a presentation of viable alternatives for decision. Third, the conference stressed that management decisions will not always wait until the 99 percent confidence level is reached—that managers must be prepared to make decisions on the basis of information available, since in the fast-moving 1970s, no decision is in itself a decision by default.

These are the pressures and feelings of urgency which bring me to your program today, and condition my remarks. First let's look at the *why* of fisheries statistics, at the biological purposes which have guided our collection of data for many decades, and at the economic and socioeconomic needs which are exerting an increasing impact on data planning for the future. Then let's consider *who* collects these data and why, and what steps are presently under way to improve our capabilities to meet management needs; and close with a look at some *goals* we might set for future action.

Why Fisheries Statistics?

Why fisheries statistics then? The primary drive for decades has been to serve biological purposes—to protect and better utilize the stocks on a fishery-by-fishery basis. Because most of you are familiar with this area, I will only enumerate briefly.

The first necessity is to understand the fundamental biology of the species; for example, age-length-weight relationships; growth characteristics and rates; catch by area; and other biological characteristics influencing the fishery.

The second obvious necessity is for continuing assessment of the impact of the fishery on the stocks; for example, catch-per-unit-of-effort as an index of population strength; cumulative catch by area within catch quota limits; and shifts in biological characteristics.

A third and more sophisticated use of data is for prediction and forecasting purposes; for example, prediction of available harvests from prerecruitment-year class strength; forecasting of environmental impacts on distribution and availability; and more complex modeling of predicted effects of changes in environmental conditions, harvest regulations, etc.

Before progressing from the relative comfort of these biological data to the comparative insecurity of information required for economic and socioeconomic purposes, let me further document the urgency for broadening our purview to include not only concern for the welfare of fish and fish habitat but concern also for the well-being of the fishing enterprise. In 1972 the National Advisory Committee on Oceans and Atmosphere (NACOA) took particular cognizance of the fact that in less than 20 years the U.S.-supplied share of fisheries products to the Nation had dropped from about 70 percent in 1955 to only about 35 percent in 1972, and that this trend showed no sign of being reversed. In consequence, NACOA called for development of a national plan for use of fishery resources, with a target of a 10 percent increase in the U.S. share of the domestic market by 1980.

NACOA's proposed steps for implementation of this national goal included determining present productivity of fishing areas of interest to the United States, projecting productivity that could be expected in 10 years if a program of ideal conservation were adopted, and relating this increased productivity to domestic market requirements.

In response to NACOA's charge, and in cooperation with the States and fisheries user groups, the National Marine Fisheries Service has developed and just this past week released a review draft of a National Plan for Marine Fisheries. The first of the four goals in that plan is the unquestioned basic requirement to support all the others: to restore, maintain, enhance, and utilize in a rational manner fisheries resources of importance to the United

States. The other three goals address particularly socioeconomic needs and benefits: to develop and maintain healthy commercial and recreational fisheries industries; to increase the supply of wholesome, economically priced seafood products to the consumer; and to improve the contribution of marine resources to recreation and other social benefits.

Throughout this National Plan for Marine Fisheries, the need for a comprehensive data base is stressed, primarily for purposes of maintaining the resource, but also with respect to meeting the other three goals of the plan. For example, one of the five recommendations of the plan requiring extensive augmentation of data-collection capabilities is designed to strengthen U.S. industry to enable it to provide increased supplies of fisheries products at competitive prices. The two subrecommendations are: (1) to establish effective fishery development programs to enable the U.S. fishing industry to enlarge its share of markets through increased productivity, lower costs, and increased acceptability of fishery products to the consumer; and (2) to design fishery management plans to permit increased industry efficiency and lower production costs.

In the context of data system needs, the first subrecommendation for fisheries development programs requires not only greatly broadened stock assessment activities, extending particularly to presently underutilized and unutilized stock, but also statistical and other studies of market and demand functions.

The second subrecommendation for fish management planning, to permit increased industry efficiency and lower production costs, addresses itself squarely to the problem of unowned resources, excess effort and overcapitalization, and the need in some fisheries for establishment of limited entry programs.

The second drive for fisheries statistics has been to serve economic purposes to enhance net economic yield. The first necessity is to maximize fleet efficiency and benefits to the fishing industry. As a statistical basis for supporting this goal, we need to gather data on: the quantity and value of catch by type of harvest, and type of gear; the characteristics of the fleet; and the landings, market values by area, and the season. The second necessity is to improve distribution and marketing for benefits to the fishing industry and the consumer.

The third drive for fisheries statistics has been for the socioeconomic benefits to individuals and the public. This requires: cost/earnings analysis for improved business management by fisheries entrepreneurs; tax collection for support of management functions—normally based on catch statistics which require a data system based on landing tickets; quantitative basis for support of positions in the political arena on domestic issues, in international negotiations, and for generation of financial support; assessment of relative values of different resource uses, such as fisheries versus other uses of habitat—dams, development of marshland, etc., and distribution of benefits—recreational/commercial harvests; and resolution of common property problems: effort management—the assessment of existing (and estimate of optimal) levels of effort, for determination of distressed status of the fishery which may mean curtailing federal subsidies for vessel construction, and may indicate possible need for effort management, and under license limitation approach, determination of eligibility for licenses depends on adequate data concerning individual history of past participation, successful fishing, degree of dependence, etc.

When NACOA in 1973 reiterated its demand for a national fisheries plan, it cited 6 conditions for development of such a plan, one of which is so pertinent as to deserve direct quotation.

> Conservation is not realistically achievable by biological management considerations alone. The federal government must also work out an approach to economic regulation of the industry with due regard for historic rights and social consequences. NACOA believes that unless there is a limit to fishing effort, the inherent surge to overcapitalization in any successful fishery will soon make it marginal. Restoration of fisheries already marginal can be brought about only by such means.

Economist Jim Crutchfield phrased the problem and the case for limitation of an effort quite succinctly in a 1965 paper on economic objectives in management: "The factor that really distinguishes fisheries (and a few other resource industries) is the inability of an individual operator to establish ownership or effective control over the basic resource. ...Whenever a resource is free it will be used to the point where its marginal yield is zero. ...In almost any fishery in which prices of the end product are high in relation to the costs of producing it, the industry is capable of destroying itself, if not biologically, then certainly in an economic sense; and unless restricted it will do precisely that."

Mr. Crutchfield went on to describe what he called the "Concept of Conservation by Inefficiency." Most fishery conservation programs have not viewed the problem as one of conserving both fish and the labor and capital used in converting the fish to useful products. They have been limited (and in some cases constrained by law) to consideration of ways to achieve the largest sustained physical yield. All too frequently, the easiest way of doing this has been to adopt methods of regulation that reduce economic efficiency—either directly, or indirectly as a side effect. There can be no doubt that fish can be "conserved" by preventing people from using the most efficient kind of gear. From the standpoint of business practice, however, curtailment of fishing effort through deliberate reduction of efficiency leaves something to be desired.

Who Collects Data and Why?

There are three main data-collection sources. The first is private enterprise—fishermen, companies, associations, unions. The reason is obvious: fishermen want to catch more fish; companies want better return on investments, a competitive market edge, etc. For example, fishermen are induced to participate voluntarily in the Coastwide Albacore Logbook program by the program returning to those who do participate a summary of previous catch concentrations by area and date. This is useful to fishermen since they collect similar data on their own concerning preferred fishing grounds, dates, etc.

The second data-collection source is research groups—universities, consultants, special teams. These special studies are usually short-term and for project purposes.

The third source is fisheries management agencies. The federal government now has primary responsibility in accumulation of regional and national statistics, particularly in market data area. However, the Great Lakes and Pacific States are heavily involved both in collection of data and in its use for management purposes. For this purpose, minimum requirements are for: landing tickets for tax collection and data accumulation; vessel registration for data on vessel type and capability and to keep track of the vessels; and special logbooks. On other coasts only limited data is collected by many States. The federal costs of data collection are more or less inversely proportional to levels of state involvement. Indeed, when the Pacific States look at their costs, they think seriously about sending the federal government a bill for reimbursement of their share in the effort—retroactive to about 1925!

There are problems of overlap and duplication in data collection. Repeated interviews and surveys by a succession of agencies antagonize those surveyed. Planning and coordination is very important to prevent unnecessary university and agency overlap.

Challenges for the Future: What Goals Should We Set?

There are three main goals to be set for the future. First is the acceptance of a broadened scope of management responsibility. This impacts upon the biological purpose of fisheries statistics—the protection of the resource and augmentation of yield. This objective must be emphasized as absolute first priority, for without a productive resource, all other goals become meaningless. The concept of optimum utilization rather than maximum sustained yield is gaining momentum in the literature, among fisheries leaders, and now in planned legislation. This acceptance of management responsibility also affects the economic purposes of statistics—the enhancement of net economic yield. We need clarification of a management mandate of responsibility for *people* as well as for *fish*—for a viable fishing industry; therefore, management must promote efficiency rather than strangulate through planned inefficiency. Permit me here to observe that the suggested legislation in my view needs a bit of augmentation in this area.

Finally the acceptance of a broadened scope of management responsibility influences the socioeconomic benefits to individuals and to society at large. It should benefit entrepreneurs through assisting them in developing sound business approaches to the fishing enterprise based on good statistics and established business principles, and, where appropriate, explore ways to remove the primary cause of management by inefficiency—excess effort in harvesting of a common property resource. Where appropriate, this can involve vesting of the right to harvest under some sort of effort management program which, of course, includes limitation of entry. It should also benefit society through: (1) collection of a reasonable fee for the right to harvest—economic rent to the rightful landlords, the society as a whole; and (2) effective documentation of issues and answers as basis for decision-making and source of public support (on domestic issues, and in international negotiations as basis for U.S. policy).

The second goal to be set for the future is the achievement of data collection standardization and compatibility—a must! One worthy example is the Gulf of Mexico Environmental Inventory. Scientists representing all Gulf States and the federal government joined together in a technical planning panel and worked under the auspices of Gulf States Marine Fisheries Commission to standardize methods of collection, degrees of precision required, and EDP data management procedures for environmental data collection for the entire Gulf of Mexico region. The Pacific Marine Fisheries Commission Coastwide Data System Study represents this effort in process.

One important aspect is that data compatibility must match the geographic extent of fishery. Particularly as we look ahead to extension of U.S. jurisdictional control over fisheries conservation and development, we must plan for institutional arrangements that will generate regional solutions to regional problems. The suggested legislation developed by the Council of State Governments is toward this purpose—and federal legislation is under drafting by congressional committees at this precise moment. Clearly our data systems must be devised to match any needs for accumulation and analysis over the total range and extent of the fishery. To that end, for example, we are seeking ways to make U.S. and Japanese albacore fishery data compatible in order to cover the full range of the species, thus bridging the North Pacific and, as we look to extended jurisdiction, we must insist on data from both domestic and foreign harvesters of our resources.

The final goal is the development of suitable institutional support and equitable funding. General fund support benefits will derive to the general public. No effort to generate special support for more or less self-sustaining operations should reduce general fund obligations for continued support in the public interest. In addition, a state/federal partnership will be required for success. State responsibilities would appear to be: (1) collection of data needed for internal management purposes; (2) achievement of prompt and accurate processing of data and dissemination of data summaries; and (3) cooperation with other States and the federal government to achieve compatibility of data, standardization of methods, etc., to facilitate input into regional and national systems.

A sport fisherman's delight—a nice catch of spotted sea trout from the Gulf of Mexico.

Federal responsibilities appear to be: (1) initiative and leadership in standardization of data needs, methods of collection and coding, achievement of compatibility, etc., and assistance to the States in achieving these goals; and (2) funding of regional and/or national systems for accumulating processing, and arraying data inputs from States, private sector, federal, and other sources, for regional, national, and international purposes.

Solid institutional commitment will be necessary for the success of this final goal. Everyone agrees to the principle that management decision should be based on sound and comprehensive data, but the step from philosophical agreement to operational implementation is a long one. There must be a solid institutional will to overcome factors of inertia, habit, inherent chauvinism, and many other human factors. In addition, the commitment of needed funds for facilitation, and some sacrifice of historic continuity, must be made in favor of future gains in relevance and effectiveness.

Recreational Fishing Interests — Conflicts and Cooperation

Dr. Frank E. Carlton
National Coalition for Marine Conservation, Inc.
Georgia

Our world is in the midst of a revolution, possibly the greatest social change since the beginning of time. Elisabeth Mann Borgese in "Pacem In Maribus" speaks of the "marine revolution" and compares it with the "long list of great disjunctures that have marked human history — the political, industrial, socioeconomic revolutions of the past, the technical and biological revolutions of the present."

The "Third Annual Report of NACOA (National Advisory Committee on Oceans and Atmosphere) to the President and the Congress" dated June 28, 1974, introduces its findings with the statement, "This year NACOA worked with the consciousness that our society may well be on the threshold of a major discontinuity in human history."

Joseph Alsop, in the *Washington Post* of January 27, 1975, quotes Secretary of State Henry Kissinger as saying, "As an historian, you would have to predict that our kind of society would very probably not last much longer. But if you are an official, you have no such freedom, you just have to do the best you can."

Sober words — "disjuncture," "discontinuity" — and sober thoughts — "that our kind of society would very probably not last much longer" — but Secretary Kissinger's conclusion, "you just have to do the best you can" implies a definite job for every man and the necessity of commitment to that specific effort. The effort we are concerned with today is a very real part of this social revolution, a very real part of the "marine revolution" and that is the subject of extended fisheries jurisdiction and the resultant opportunity for comprehensive natural resource management in the sea.

Fisheries Conflicts

Successful management must anticipate user conflicts and enforce dispute settlement mechanisms. The long history of problems between sport and commercial fishermen offers ample testimony to this point. Fisheries conflicts are associated with competition for declining stocks and disputes regarding areas, access, gear, habitat destruction, traditional rights, and legal prerogatives.

Fisheries conflicts are caused — if a single cause can be pointed to — by a behavioral preference for aggressive self-assertion and violence, rather than rational, mutually beneficial compromise. This particular quality is frequently associated with a lack of information and failure of communication.

The importance of recreational fishing versus commercial fishing has been variously compared in many ways over the years. Many of these items can be grouped into basic categories and viewed under three purposes for comparison.

Recreational	*versus*	*Commercial*
1. Social luxuries — expressed as aesthetic pleasure or frivolous recreation.		1. Economic necessities — expressed as employment and/or nutrition.
2. Conservation — expressed as an ethical principle.		2. Exploitation — expressed as material profit.
3. One big business.		3. Another big business.

These comparisons have been alternatively extolled and/or criticized as reasons why one group or the other should be accorded a preferential right to a particular fish or area; but there are few examples of reasonable and timely resolutions to these conflicts, either by fishermen themselves or through a "third-party" mediating agency. Both sides, recreational and commercial alike, have pointed an accusing finger at alleged government bias, and there has been some. Recreational fishery/commercial conflicts are often resolved politically by administrative decision or voter preference and all too frequently this solution has been neither fair nor rational.

The most significant comparison between sport and commercial fisheries has not been made, and cannot be made, yet, because it requires a study of the economic value of the sportfishing industry. Present estimates indicate that there are more than 34 million fishermen who spend over $5 billion annually. The most accurate direct comparison of sport and commercial fisheries was done by Gruen and Gruen of San Francisco, California. In a study entitled, "A Socio-Economic Analysis of California's Sport and Commercial Fishing Industries" (June 1972) they stated that the net economic value of commercial to sport fishing was $43,356,533 compared to $201,120,000. These figures substantiate the travesty of viewing the sport fisheries merely as a frivolous recreation, but administrative traditions and emotional bias die hard. In addition, there are other significant limitations to achieving a solution to fisheries conflicts.

First, the physical perspectives of the fishermen themselves and, second, the socioeconomic perspective of the government administrator. Physical limitations — the invisibility of the resource and the temporal camouflage of lengthy and complex life cycles — have aided and abetted the avoidance of a commitment to a conservation ethic by recreational and commercial fishermen alike. In a similar sense, the lack of administrative perception of social aspects of recreation/commercial squabbles and, even more important, the administrative failure to anticipate developing issues and establish a means of rational negotiation prior to achieving a point of spontaneous combustion, have contributed an equally serious set of difficulties. All of these factors, as contributions to fisheries conflicts, have a parallel significance as specifying needed objectives for management.

Let us further consider the anatomy of recreational fisheries conflicts for the moment.

Lessons in History

There are literally hundreds of examples of fisheries conflicts which could be listed chronologically from at least the 15th century to the present. What can be learned, perhaps simply relearned, from yet another list of already familiar examples? Probably very little. One general comment does seem pertinent. Analyzing such a list in the light of other considerations about the history of fisheries science supports the assertion that our more serious problems are related to a specific lack of management rather than to an attempted program which was simply ill advised or not sufficiently funded.

Three current recreational/commercial conflict cases — Cape Hatteras, the Judge Boldt decision, and the present striped bass problem — may offer clear suggestions toward certain management objectives for the future.

The Hatteras and Boldt examples are important because of the dramatic illustration of the harm that legislation in itself can do when the original intent of the law is circumvented by evolving technology and other societal changes. Haul seine methods and out-of-state commercial fishing and recreational use in 1937 are in no way comparable to the total impact of pressures on migratory striper stocks at Hatteras today. *Tempus fugit* — the lesson from the present Hatteras conflict is very clear.

The original Indian treaty affecting the fishing rights of some 18 northwest tribes had a similar intention to that of Cape Hatteras National Seashore Act: to provide for a continuation of traditional practices. Several facets of the present situation illustrate the inequity of purely legal solutions to complex natural resource problems. It may be "legal" but it is certainly not "just" to allow the treaty Indians on the Nisqually River to catch a greater number of steelhead in 1975 than the total natural run was at the time the treaty was signed, especially in view of the fact that the enhancement of the run has been funded entirely through sportsmen's dollars. Again the lesson is very clear.

The northeast Atlantic striped bass problem is somewhat different in that most mid-Atlantic and New England state laws were written in timely fashion on the basis of good biological data. But again, that timely and biological data was restricted to local conditions. For some years it has been evident that the striper is a regional as well as a local resource and that regional conflicts on regulations and enforcement — on size, catch, and season — are generally harmful to everyone concerned. Once again an obvious problem and an evident solution.

Trolling offshore for sailfish.

These three examples lead to at least two fundamental conclusions regarding the precipitation of fisheries conflicts and related objectives for management.

1. Conflicts are characterized by a definite temporal component; therefore, the first objective of management must be *timeliness,* and

2. Any solution to conflict is limited by that same temporal quality; therefore, solutions must include the *capacity for change* — to reflect better data as it becomes available and to respond to socioeconomic evolution.

These two simplistic criteria appear self-evident to the point that the unwary may take their accomplishments for granted. Not so. Although these qualities are simple, self-evident, and have been obvious for hundreds of years, we still have obvious problems in accomplishing their practice. One of the principal difficulties today is inflicted by an antiquated and obsolete vision of science expressed most eloquently by Ms. Borgese in her thoughts on peace in the sea.

> Most people agree that old intergovernmental formula of science administration has run its course — that, reflecting a 19th century concept of science as passably observing and describing nature, it has lost its efficacy. This (view) of science leads on the one hand to catastrophic gaps between knowledge and action, as was the case when scientists fully well knew what was happening to the blue whale but politicians failed to act on that knowledge and brought this marvelous beast past the edge of extinction. On the other hand it engenders duplications of effort that degenerate into waste. There are today at least 13 intergovernmental agencies and 14 committees at the world level that are dealing with matters on science and technology within the U.N. system and sometimes their frames of reference are literally the same.
>
> Experts point out these agencies are essentially vertical organizations surviving from the needs of government activities as seen at the end of the last century. There is no concept of a multi-disciplinary horizontal systems approach to activities. It seems unlikely that governments will solve this problem in the near future. It can be solved either by a total crisis or a group of people putting forth really new ideas. The new concept of science, as not only observing, but changing nature must be embodied in a link between research and action. This requires institutional innovation to enable scientists to participate in the decision making process of planning and government.

It is evident that this 19th century view of science as a passive observer must evolve into both a more immediate awareness of current needs and also into a greater capacity to act.

A second problem we have today is related to the increased data needs implied by future management requirements. We are on the brink of a great revolution requiring administrative decisions (management) which will obviously demand a great increase in data collection and storage. I fear we will face a succession of decisions — successive moments of decision — and be swimming in literally oceans of data, only to drown through the lack of a mechanism to select priorities and make decisions. It is entirely possible to see extended jurisdiction and its related management institutions manifested as endless stacks of IBM cards and miles of magnetic tape will never bring to life a viable management program.

I regret this criticism, but we can no longer afford a scientist who wishes to cloister himself with the myth of objectivity — particularly as an excuse to avoid commitment.

We live in a troubled and complicated world. We must have sound decisions, better data, and better science to merely survive. I reiterate my simplistic fundamentals for management: timeliness and capacity for change.

Recreational Fisheries Management

Now let us discuss specific management requirements from the recreational fisheries perspective. The National Coalition for Marine Conservation has worked diligently to develop its own version of a management bill which will shortly be introduced into the Congress of the United States. The fundamental provisions of that bill include:

1. Maintenance of maximum authority at the local and regional level;
2. Balanced input from all users, recreational and commercial alike;
3. A system of administrative checks and balances whereby the fishery cannot be dominated by any particular user group or on the other hand by the federal government;
4. A dispute settlement mechanism which is relatively quick and as politically insulated as possible.

The broad principles these provisions reflect have also been expressed. A draft version of the following statement is being circulated nationally to recreational fishing and conservation organizations for their comment and endorsement.

Statement of Basic Principles and Provisions

The following purposes and provisions are held by the undersigned recreational fishing and conservation organizations to be the *minimum* basic requirements for a marine fisheries management regime:

1) A clear and unequivocal commitment to long-term conservation goals including the restoration of depleted stocks and the maintenance of productivity of all fisheries.

2) A clear and unequivocal commitment to obtaining and maintaining the scientific data base essential to effective fisheries management and to expansion of biological research concerning the interdependencies between species, the impact of pollution, and other factors bearing upon the abundance and availability of commercial and recreational fish species.

3) A clear and unequivocal commitment to the broad concept of optimum yield in management of the fisheries in place of the narrow concept of maximum sustained yield, i.e., to consideration of recreational, social, ecological and economic as well as biological factors in the determination of allowable catches within every fishery under management.

4) An opportunity for substantial participation by all parties interested in the fisheries including State administrators, commercial, recreational, conservational and other interests at every level of policy making and regulation making in the fisheries management procedure.

5) A clear and unequivocal commitment to equitable allocation of the allowable catch in each fishery under management with due regard to the interests of recreational fishermen in the fishery itself, or in other species related to, or affected by, the condition of such fishery.

With the advent of extended jurisdiction and comprehensive

management, the immediate issues for decision will be:

(1) Establishment of exploitation levels for stocks under U.S. jurisdiction;

(2) Allocation; and

(3) Foreign participation.

At least 30 stocks of interest to U.S. fishermen are at critical levels of exploitation now. The third draft of the National Fisheries Plan for marine fisheries, dated June 1975, lists 14 species as overfished and 16 as fully utilized. Allocation of some species between different types of American fishermen, between commercial/recreational users generally, and between declining foreign and increasing U.S. interests should be made as soon as jurisdiction is established. A legally mandated, scientifically capable institution must be created to do this work. The required institution does not now exist.

A Saltwater License

But the immediacy of these issues is predominantly a federal problem. What relevant point on management should be considered from the state perspective? Is there an immediate problem which will confront the state administrator in the near future? In 19 out of 22 coastal States there definitely is.

There is absolutely no way any state agency responsible for marine fisheries management can do its job without knowing how many fishermen there are and what they catch. All 22 coastal States count commercial fishermen through individual and boat licenses and keep records on their catch, but only three States, Alaska, Washington and Texas, license and therefore accurately count their recreational fishermen. No general catch statistics are kept although excellent data on certain special species has been obtained for years. California requires a license to fish in saltwater, but the mechanism involved does not allow a specific count of recreational anglers to be determined. Alabama and Louisiana also have a token marine license, but it is not enforced and therefore does not provide a basis for a meaningful management program or supply statistical information.

A brief history of the saltwater license question is appropriate. In 1956 the Sportfishing Institute (SFI) collaborated with the National Wildlife Foundation in the first national survey on a saltwater management program. The SFI Bulletin (#46 for July 1973) reported that the survey expressed an "overwhelming endorsement of the saltwater license as a device to raise funds." Resolutions endorsing a saltwater license as the most feasible method for raising funds for needed recreational fishing conservation programs were adopted in:

1957 — by the National Wildlife Foundation
1959 — Izaak Walton League of America
1961 — Sportfishing Institute.

The Sportfishing Institute resolution followed completion of a survey which included 105,000 saltwater anglers. The survey clearly demonstrated that saltwater anglers

(1) Favored development of recreational fishing conservation program by state agencies — 93 percent;

(2) Believed research to be an integral part of such programs — 94 percent; and

(3) Desired improved physical access — 93 percent.

The SFI study concluded with two basic recommendations:

(1) Funds must be tied to definite conservation programs; and

(2) Funds from recreational sources must be devoted to recreational programs.

In 1961 Richard H. Stroud delivered a paper entitled "Some Marine Sportfishery Problems" to the Sixth International Gamefish Research Conference. Mr. Stroud stated:

> Experience with inland fishing license shows that an annual state fee of about $3 is needed to start. It also shows that administration information, education and law enforcement collectively would require about 25% of the funds in the early program stages leaving 75% for development and applied research. Funds should be safeguarded against diversion by designing politicians in non-related programs. Licensing should be acceptable and saleable if a budgeted program is presented and if there is a policy making body or commission representing the anglers to oversee the use of the funds. Yield collectively (1961!) would be over $18 million annually at present and an average of over $800,000 per state. The amount of money would provide $600,000 for development and applied research and $200,000 for necessary administration.

In 1967 the United States Department of the Interior published a "Report of the Committee on Conflicts between Sports/Commercial Fishermen." Excerpts from that document state:

> The service recognizes the vital role of a vigorous and successful commercial fishing industry in the nation's economy and the high value of recreational fisheries. The Bureau of Commercial Fisheries and the Bureau of Sportfisheries and Wildlife must each develop public awareness of these multiple values so that conflicts may be solved on the basis of sound biological, economic and social values. The service recognizes that each state has an important responsibility for unbiased and positive leadership in resolving conflict.

Later in the report, specific recommendations are made to state fishery agencies:

(1) Appoint a top official to investigate sport/commercial conflicts;

(2) Improve media relations in reporting;

(3) Promote sound and uniform fisheries laws;

(4) Initiate a recreational license for marine and Great Lakes;

(5) Cooperate with States to develop uniform laws and eliminate interstate conflicts; and

(6) Maintain better administrative and regulatory control over fishery resources.

Please note that the prime reason for advocating a fisheries program that would include an accurate determination of the numbers of recreational anglers and their catch — inland and marine alike — is based on the information needed for management. This should be considered a minimum requirement. The funds which could be generated are obviously important and supplementary to any recreational fisheries conservation program, but not primary. Direct income from license fees and Dingell-Johnson funds would constitute a lot of money now — about $1 million per State annually — but the expenditure of those funds without programs based on sufficient data would be farcical.

Many recognized authorities, including Richard Stroud, Henry Lyman, Frank Moss, and others, have suggested that angler education and the resulting demands for research and physcial improvements are the proper route for securing the political support needed to obtain a saltwater fishing license. But if the SFI survey is accurate, the support is already there. The national survey (1961) clearly demonstrated that the vast majority of anglers supported the principle of state marine management programs. Obviously there is opposition but these statistics suggest that the dissension comes from less than 10 percent of those who fish in the sea.

Another interesting aspect to the question of angler support is revealed by a comparison of the percentage of pure ocean anglers to the total of inland and marine combined. The 1970 National Survey of Fishing and Hunting and several state estimates of angler numbers indicate that less than 25 percent of the total angling population fishes exclusively in the ocean; therefore, requiring one license to fish in all state waters would cost 75 percent of recreational fishermen nothing beyond what they are already paying!

A Single Comprehensive License

The plan I personally recommend to state agencies includes:

(1) Expeditious establishment of a Marine Fisheries Advisory Council composed of commercial and recreational fishermen and appropriate scientists and administrators.

(2) Initial use of the council for angler education programs, including general conservation, boating safety, fishing techniques, and basic fisheries research needs. This educational phase should be specifically designed toward equipping the council to be able to review policy, oversee the use of state funds, and adjudicate user disputes.

(3) The planning of a comprehensive coastal research program which would include a specific recreational fisheries division. Activities within the recreational fisheries division should include:

(a) specific marine gamefish research programs including stock assessments, life studies, food chain interrelationships, etc.;

(b) water quality baselines and monitoring systems;

(c) estuarine survey and protection programs;

(d) physical improvement programs, including water access, dry storage, catwalks, artificial reefs, and wetland and estuarine sanctuaries; and

(e) a single comprehensive recreational fishing license covering all state waters, fresh and salt.

(4) License legislation which would guarantee that a comparable percentage of funds generated from recreational fisheries sources would be restricted to recreational fisheries programs. Other criteria should provide:

(a) that no catch could be sold (without an appropriate commercial license);

(b) reciprocity with States sharing contiguous marine borders;

(c) one price to residents and noncontiguous out-of-state anglers who are licensed in their State, and higher prices to nonlicensed anglers and foreigners;

(d) additional special stamp or punch card, funds rigidly restricted to that specific special project; and

(e) that a future testing requirement as a prerequisite to licensing, that is, an examination on basic conservation and boating safety (similar to a driver's license exam), would be required prior to the angler's qualification to purchase a license.

Conclusion

The latter portions of this paper dealing with specific state marine problems are punctuated with a historical perspective of over 20 years. It is important to appreciate this 20-year period in terms of the continued decline in stocks, continued increases in fishing pressure, and the additional burden of escalating development of other coastal and offshore resources. In order to understand these subjects from a more accurate perspective I would like to conclude by quoting a paragraph from an article entitled "The National Protection for Denizens for the Seas and Waterways" by Bushrod W. James as quoted from the *Bulletin of the United States Fish Commission,* Volume XVII, 1897 — some 78 years ago.

> Many wise individuals today deplore the dilatory attention to national interest that has resulted in comparative extinction of many really valuable creatures whose abundance seemed but a few years ago to be inexhaustible. Should not everyone energetically lend his voice and influence to prevent further loss to both individual and government? A war of extermination of the human inhabitants of the remote corners of the country would justly be considered a heinous, cruel outrage; but is not the destruction of lower animal life in vast multitudes equally cruel? If mankind has its sources of life's necessities cut off they pine and die. Thus we as a Congress (The National Fishery Congress of 1898) should urge full legal protection through both home and international law, for the food fish upon which vast numbers of human beings depend for all that makes life comfortable; while in some places neglect to pass such laws actually results in suffering and death. We do not deem it right to propose protection only, but should follow the proposition up by active earnest work for the desired result.

Management of Marine Fisheries Resources

David H. Wallace
Associate Administrator for Marine Resources
National Oceanic and Atmospheric Administration
U.S. Department of Commerce

There has been a growing conviction that existing institutions, both domestic and international, have not succeeded in conserving fisheries resources in an effective manner. I have just returned from Edinburgh, Scotland, where I spent two weeks attempting to achieve some degree of rational conservation and management of the fisheries off our Northeast Atlantic Coast in which 16 other governments are, to a greater or lesser degree, involved.

Insofar as the International Commission for the Northwest Atlantic Fisheries (ICNAF) is concerned, it is perhaps the most sophisticated of our international fisheries regimes. Yet, despite some accomplishments and the determined efforts on the part of the United States and Canada, ICNAF has been unable to prevent serious overfishing in the northwestern Atlantic Ocean, and it has been unable to resolve the serious conflicts between our fishermen and those of the various nations. Insofar as I am concerned, as the United States federal commissioner to ICNAF and the head of the delegation to that critical meeting, it is clear that the distant water nations fishing off our coast are unwilling to impose and accept the controls on fishing needed to restore within a reasonable time the biomass which has been severely reduced by overfishing. This assessment of the damage to the stocks is unchallenged, and the course which must be followed to correct the condition had been agreed upon by all participants. Yet at this last stage, where it was necessary to set the so-called second tier quota at a level which would halt the decline and bring about the restoration of the stocks in a reasonable time, the vote was Canada and the United States against all the others.

This is neither the time nor place to have a detailed analysis of what happened or the implications. I am certain that the countries who were aggressive in getting the quota raised to a level unacceptable to the United States and which would not result in even starting to have recovery of the biomass in eight to 10 years will rationalize their actions as desirable and necessary and not really damaging to the stocks. The others who voted with them will agree and support this position. I cannot accept this view and in the meetings so advised the other countries. The failure of ICNAF to accept the responsibility of rebuilding fisheries stocks will obviously have an impact on the Administration position on extended jurisdiction and on Congress in relation to pending legislation. In my opinion we must act decisively, and I can assure you that both the National Oceanic and Atmospheric Administration (NOAA) and the National Marine Fisheries Service are deeply involved in the discussions within the Administration as to what our position should be. We should have a decision on this soon.

The 200-Mile Limit

I am convinced that sometime in 1976 the United States will move into extended fisheries jurisdiction out to 200 miles either by the Law of the Sea (LOS) Conference or by congressional action. I must say at this point just two days before leaving for Edinburgh I testified before the Senate Committee on Commerce regarding proposed legislation to extend the fisheries management responsibility and authority of the United States out to 200 miles.

Under a 200-mile regime, the United States will control the largest and most valuable fisheries resources available to any nation in the world. Coupled with an effective management program, it will give us the opportunity to establish sound management programs over our coastal fish and shellfish that will assure conservation of stocks and permit efficient methods of utilization. This control of fisheries stocks is essential no matter how extended jurisdiction is attained. The present Administration's view is that it should be attained through multilateral agreement at the LOS Conference and not by unilateral action and international confrontation. All of us are quite aware of the impatience and the pressures in the Congress and in the domestic coastal fishing industry for a unilateral declaration of a 200-mile fisheries zone now. This entire matter is under intensive review within the executive branch, but it is quite possible that the failure of the ICNAF nations to vote the necessary controls which we felt were essential to conserve the stocks may be the situation which will trigger action by the Congress regardless of the views of the Administration.

There can be no question that at least some of our important fisheries resources are in trouble and show serious depletion. Much, but not all, of this decline is a result of excessive foreign effort off our coasts and the failure of existing international agreements to protect these resources. There is a feeling that problems with the coastal fisheries come about from foreign fishing alone, but this is not correct. Domestic overfishing has also been damaging in some circumstances. Conservation of fisheries resources requires an effective management regime which will control both foreign and domestic fishermen.

The 200-mile exclusive economic zone will, in our estimation, provide the needed framework for conservation of our coastal fisheries resources. But a 200-mile exclusive zone alone is not enough: it must be coupled to a management regime with new mechanisms which can effectively regulate harvesting of fish by foreign and domestic fishermen. Our objective is to protect the fisheries resources to insure their survival at optimum levels.

Components of Effective Management

NOAA has been studying the implication for effective management of fisheries in a 200-mile zone and the kind of legislative authority required to provide the necessary management tools. Based on this study and the comments and discussions we have had with state and industry leaders and others, we have identified three major components of what we consider to be an effective management regime.

First, the regime must have a policy-determining and regulating component. This component must consider local, state, national, and international issues; it must be decisive and equitable in the decisions it makes because such decisions will affect people and how they live.

Second, the regime must have a system for data collection and analysis. Catch data, supplemented by resource surveys and a continuing program of fisheries research, will be the basis for assessing the condition of the resources and the effects of fishing on the stocks.

Third, the regime must have means to enforce the regulations and adjudicate violations at state, national, and international levels.

Management of stocks which migrate across jurisdictional boundaries requires a system of strong regional fisheries organizations. This concept is supported by almost all of those who have commented on the extended jurisdiction study. The problems may be local, state, national, or international in scope, but their solutions are best developed in the region by those most intimately concerned. For these reasons, regional marine fisheries councils should be established to develop the management plans for each of the fisheries in a region. The three existing marine fisheries commissions could serve in an advisory role to at least some of the councils. But it is NOAA's view, and that shared by a number of those who reviewed the report, that the commissions should not be the regional fisheries management organizations. There are varying views on the composition of the councils. It is our current view that they should be composed of representatives of the state agencies responsible for management of marine fisheries, along with appropriate representatives of the federal government. This is where management expertise lies. In addition, some formal mechanism for obtaining advice in the very beginning of the planning process from all concerned groups must be part of the system.

Facts support our position that the States must have a strong role in the development and implementation of management plans. States already have a capability for management which must be utilized insofar as feasible and practical.

The domestic harvest within three miles of shore is mostly comprised of stocks that migrate across the boundary of the three-mile territorial sea or the boundaries of adjacent States. There must be a single focus for the development of management plans for each stock throughout its range. This central focus can be vested in the regional councils, with implementation by appropriate state and federal authorities. For those fish stocks that live primarily outside state jurisdiction, the federal government must have ultimate management responsibilities, for no other reason than that the other players are foreign nationals. For fish stocks which occur predominantly within the jurisdiction of a single State (i.e., within three miles of the coastline) the management responsibility should remain with the individual States. For fish stocks which live within three miles of the coastline but move across state boundaries, some form of regional management must be adopted. The State-Federal Fisheries Management Program should provide a solid foundation for this type of cooperation.

The creation of a 200-mile economic zone will give to the United States preferential rights to coastal fisheries resources. We must create the situation which will permit United States fishermen to take up to their potential within the limits of conservation. If our fishermen, commercial and recreational, can properly utilize the stocks, foreign fishing must be halted.

International Aspects

It is my personal view that under extended fisheries jurisdiction, foreign fishermen must pay for the privilege of participation in our coastal fisheries. This should include the fair share of all the costs of management (administration, research, and enforcement) and a fee for resource use as well. Additionally, as a condition for their participation, they must provide all catch information required by the management regime as well as be subject to all other management regulations imposed by the United States. The whole question of enforcement and surveillance is a complex one, particularly in determining how much is enough and what is the best combination of methods to use. We are currently working very closely with the Coast Guard and other federal agencies on a thorough analysis of the entire problem.

So far I have been talking about the management of our coastal stocks and have not commented specifically on the management of anadromous species or the highly migratory species such as tuna that cross the 200-mile lines. The United States' position in the Law of the Sea Conference has been that the basic responsibility for management of anadromous species must be lodged in the country where the fish originate. Futhermore, the fishing should be primarily confined to the economic fishing zone. However, it is obvious that certain historical fisheries on anadromous species have existed on the high seas and elsewhere. Provision must be made for appropriate international arrangements which would provide for dealing with these fisheries, in the interests of conservation and the domestic fishermen.

Regarding highly migratory species, such as tuna, our position has been that they must be managed in the interest of conservation and equitable sharing of the stocks by international bodies, with authority to make adequate regulations to accomplish these goals.

Conclusion

To sum up, good fisheries management needs to be an evolving process

Landing a bluefin off the Atlantic coast.

that concerns itself with the assessment, protection, allocation, and utilization of living aquatic resources in a manner that provides the greatest benefits to society. I am well aware of the complexity of the problem we face in the management of our coastal fish resources. The fish stocks are many and varied. Provision must be made for the development of fisheries management plans, each tailored to specific needs of regional fisheries problems, and prepared cooperatively with the States with advice and input from affected local interests. The federal government must hold a position of general leadership and authority for regulating those fisheries beyond state territorial waters, but it must also be exercised in concert with the state governments.

Before closing I would like to complement the Council of State Governments' Task Force on Effective State Marine Fisheries Management Programs for preparing the draft legislation that was presented at this meeting. Ideally, the States involved should get together for joint management, but to date effective interstate action has been most difficult. As most of you know, much of the difficulty results from the lack of uniform legislation which would enable the States to function effectively in interstate, or state-federal management programs, such as would be envisioned under the regional council concept. I consider the proposed suggested state legislation a major step forward in ameliorating this situation and in improving and developing effective fisheries management programs, not only among the coastal States, but also between the States and the federal government.

Picture Sources

Louisiana Wild Life and Fisheries Commission	p. 6 (top)
National Marine Fisheries Service, National Oceanic and Atmospheric Administration, U.S. Department of Commerce	p. 6 (bottom)
National Marine Fisheries Service, National Oceanic and Atmospheric Administration, U.S. Department of Commerce	p. 12
National Marine Fisheries Service, National Oceanic and Atmospheric Administration, U.S. Department of Commerce	p. 16
Bureau of Commercial Fisheries, U.S. Department of Commerce	p. 19
Fish Commission, State of Oregon	p. 44
Division of Marine Fisheries, Department of Natural and Economic Resources, State of North Carolina	p. 47 (top)
Division of Marine Resources, Department of Natural Resources, State of Florida	p. 47 (bottom)
Division of Marine Fisheries, Department of Natural and Economic Resources, State of North Carolina	p. 54
Division of Marine Fisheries, Department of Natural and Economic Resources, State of North Carolina	p. 62
Louisiana Wild Life and Fisheries Commission	p. 70
Division of Marine Resources, Department of Natural Resources, State of Florida	p. 75
Bureau of Commercial Fisheries, U.S. Department of Commerce	p. 86